Field Guide to the Ferns

and Other Pteridophytes of Georgia

Lloyd H. Snyder, Jr.
James G. Bruce

The University of Georgia Press
Athens and London

Designed by Sandra Strother Hudson
Set in Linotron 202 Trump Medieval

The paper in this book meets the guidelines for permanence and durability of the Committee on Production Guidelines for Book Longevity of the Council on Library Resources.

Printed in the United States of America
90 89 88 87 86 5 4 3 2 1

Library of Congress Cataloging in Publication Data

Snyder, Lloyd H.
Field guide to the ferns and other pteridophytes of Georgia.

Bibliography: p.
Includes index.
1. Ferns—Georgia—Identification. 2. Pteridophyta—Georgia—Identification. I Bruce, James Garnett. II. Title.
QK525.5.G4S69 1986 587'.09758
85–24652
ISBN 0-8203-0838-2 (alk. paper)
ISBN 0-8203-0847-1 (pbk.: alk. paper)

To

Rogers McVaugh and Joseph H. Pyron,
who wrote the first book dealing
solely with the pteridophytes of
Georgia and to whom this present
work is indebted,

and to

Warren H. Wagner, Jr.,
whose encouragement and constructive
comments made the
reality of this book possible.

Contents

Acknowledgments

It would be impossible to thank individually all the persons who have helped, directly or indirectly, in the preparation of this book. We do wish to thank friends and members of the Georgia Botanical Society and the American Fern Society, and staff members of the Georgia Department of Natural Resources and the Chattahoochee River National Recreational Area.

Samuel B. Jones, Jr., and Nancy C. Coile of the University of Georgia, Wayne R. Faircloth of Valdosta State College, and Donald J. Drapalik of Georgia Southern College all helped in checking and listing herbaria records. A. Murray Evans and Michael I. Cousens assisted in identifying numerous specimens and furnished information about several species. William R. Buck and Joseph M. Beitel were of particular help in resolving problems in *Selaginella.* David L. Emory was not only a welcome companion on numerous field trips, but also gave many helpful suggestions.

Thanks are especially due Warren H. Wagner, Jr., who read and checked the manuscript in its entirety and made beneficial corrections and recommendations throughout.

We are grateful to Annice Snyder and Marguerite Bruce for their company on many field trips and for their patience and encouragement while we were writing the book.

Many of the illustrations were done by Joseph H. Pyron and are from McVaugh and Pyron's *Ferns of Georgia.* The others are by Cecile Smith, a graduate student at the University of Georgia.

Introduction

What Is a Fern?

Ferns and fern allies, or pteridophytes, are among the oldest of our land plants, having evolved some 400 million years ago. They reproduce by spores and have an organized vascular system to carry water and other necessary materials. As ferns have neither flowers nor seeds, fern reproduction remained a mystery until the mid-1800s. In fact it was believed that fern seeds were invisible, and if one could obtain them, he too would become invisible. This idea found expression in Shakespeare's *King Henry IV:* "We have the receipt of fern-seed, we walk invisible."

Comparatively speaking, ferns and other pteridophytes are a small group of plants. Worldwide there are about 10,000 species of ferns and 1,000 other pteridophytes. In the United States about 400 different species have been discovered, and currently 119 known species, of which 83 are true ferns, have been found in Georgia.

Ferns and other pteridophytes are defined as "vascular plants that are reproduced by spores." Ferns have *megaphylls,* large leaves with a branching vein system, while the other living pteridophytes have *microphylls,* small scalelike leaves with only a midvein. Spores were not discovered until 1669, and the life history of ferns was not understood until explained by the German botanist Wilhelm Hofmeister in 1841.

The life cycle of pteridophytes is unusual. When spores mature the sporangia burst open, scattering the spores. A germinating spore produces a *gametophyte,* or *prothallus*—a small, usually green, heart-shaped body about one-quarter inch in diameter and attached to the soil by many hairlike rhizoids. The gametophyte bears the sex organs—the female archegonium with its single egg and the male antheridium with its numerous sperm. In a film of water the sperm swims to and fertilizes the egg. From this union grows the new fern or *sporophyte* (a plant producing spores).

Fern spores are minute bodies, smaller than grains of pepper, and usually black, brown, or yellow. They are contained in a case called a *sporangium.* Normally there are 64 spores in a sporangium, although in some species there may be as many as 500 or more. Many sporangia are grouped together to form a *sorus.* Sori are usually located on the underside of fertile leaves and to the uninitiated are often believed to be some form of insect life or disease harmful to the plant. Sori may be arranged on or near the leaf margin, scattered over the surface, in rows along the midvein, or on the edges of the leaves.

On many ferns the sorus is covered by a thin, membranous structure, or *indusium.* This structure varies with different species and may be circular, kidney-shaped, or elongate. It may be attached to the leaf by a central stalk or along its side. In some ferns with the sori along the leaf edge, the leaf margin inrolls to form a false indusium. The location of the sori and the shape of the indusia are often excellent means to identify a fern.

Sporangia are found on the fern's fertile leaves. Ferns may have both fertile and sterile leaves, which are often identical except for the presence of the sori on the fertile ones. In other cases the two kinds of leaves are either slightly or completely different, in which case the fern is said to be *dimorphic.*

FERN PARTS

Stem (rootstock, rhizome): May grow erect or horizontally on or below the ground's surface. May be short- or long-creeping, branching or not. Usually covered with scales or hairs.

Fern Parts

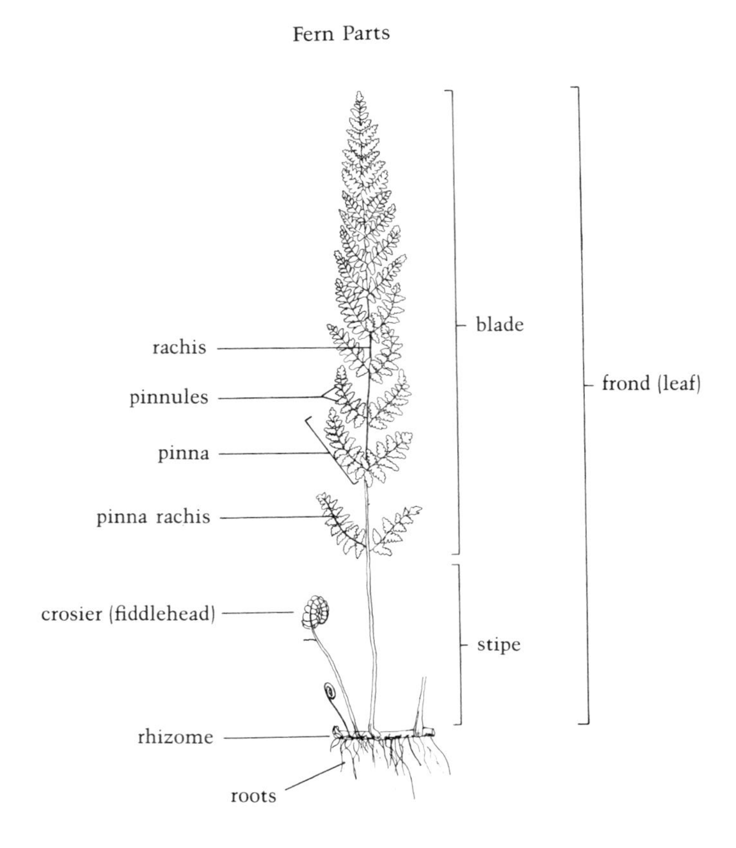

Roots: Wiry, hairlike, adventitious structures growing from the stem.
Stipe (stalk, stem, petiole): The leaf support growing from the stem. May be glabrous or have varying amounts of hairs and scales. As seen in cross section, it may be round or flat, with or without a groove.
Blade: The flat, green, expanded part of the leaf, varying in size and shape according to the species. May be simple and undivided, or cut in various degrees.
Frond (leaf): The whole organ, including the stipe and the blade. Fertile and sterile leaves may be alike or dimorphic.
Rachis (axis): The part of the stipe continuing into the blade and carrying the pinnae, or leaflets. The pinna and pinnule rachises are also known as midribs.
Pinna (leaflet): The primary division of a compound leaf blade.
Pinnule (segment, sub-leaflet): The secondary division of a compound blade.

Two other features of a fern that are important and often helpful in identifying a species are:
Veins: In most of the ferns in this book the veins run free from the midvein to the margins without uniting. They may or may not fork. In some species they unite to form a network of areoles.
Vascular Bundles: If a cross section of the stipe is taken, the conducting tissues or vascular bundles can be observed. These vary in appearance and number.

A distinctive feature of most ferns is the way in which the fern leaf grows. The tip of the growing fern leaf is tightly coiled like a watch spring. The immature fern frond is called a *crozier* (from the head of a bishop's staff), or a *fiddlehead* (like the narrow end of a violin). The fern leaf matures from the base to the tip as this coil gradually unrolls. Fern fiddleheads are an attractive sight in swamps and woodlands in springtime.

BLADE DISSECTION

Fern blades are cut or dissected in a number of different ways, so that ferns may or may not have a "ferny" appearance.
Simple: The blade is uncut with a solid form.
Pinnatifid: The blade is divided into a number of segments, but the cuts are only part way or nearly to the rachis. This can be thought of as a "part cut."
Pinnate: The blade is divided into a number of pinnae, with the cut going all the way to the rachis, and the segments stalked or contracted at the base—a "full cut."
Pinnate-pinnatifid: The blade is divided into a number of pinnae cut all the way to the rachis. Then the pinnae are divided into segments with the cut going only part way to the pinnae rachis—"a full and a part cut."
Bipinnatifid: The blade is divided into a number of segments that are further

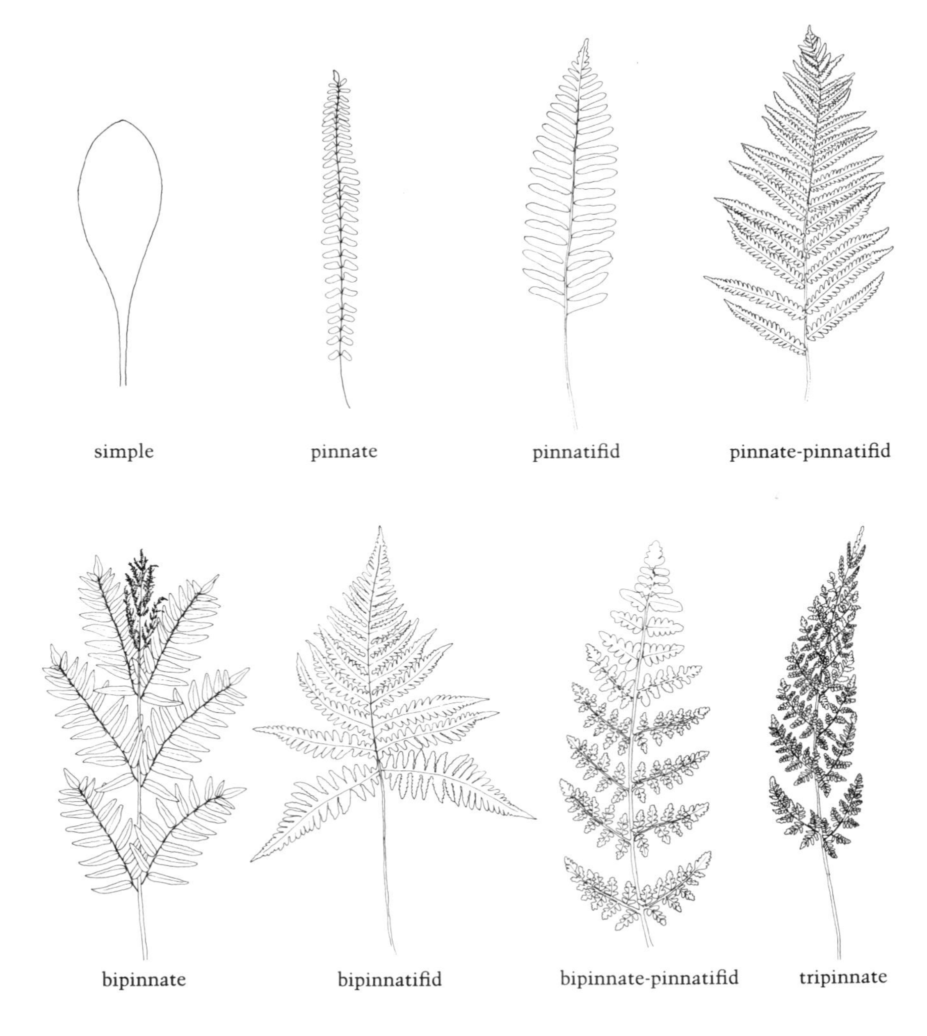

divided. In both cases the cuts are only part way to the rachis or midvein—"two part cuts."

Bipinnate: Both the blade and pinnae are cut all the way to the rachis and midvein into segments contracted at the base—"two full cuts."

Tripinnate: Blade, pinnae, and pinnules are cut all the way to the rachis and midveins—"three full cuts."

Bipinnate-pinnatifid: Blade and pinnae are cut all the way to the rachis and midveins and the segments are only part way cut—"two full and a part cut."

FERN NAMES

Since the beginning of civilization people everywhere have given plants common names as a means of communicating with each other about them. In the course of time different persons have given the same plant different names or the same name to different plants. In 1753 the Swedish naturalist Carl von Linné (known by his Latin name Linnaeus) published his monumental *Species Plantarum.* In this work he established the binomial system of nomenclature, giving each plant a generic name plus a specific epithet. The scientific names were Latin or Greek so that they could be recognized by botanists worldwide. The genus (plural "genera") is a group name and together with the epithet distinguishes a particular plant in that group. The epithet is often erroneously called the species, when in fact the species name is both the genus and a specific epithet combined. The epithet may refer to a species habitat or location, to some descriptive characteristic, or to someone's name.

Today plant names follow the *International Code of Botanical Nomenclature.* Generic names are capitalized and epithets are written in lower case with the describer's name following. Names may change under specific circumstances. Someone may later find that a name was earlier given to the same plant, and according to the rules, the first name takes precedence. Or the name may already be in use for another plant. Often it is found that the plant should be in another genus. When a name is changed, the original describer's name is placed in parentheses followed by that of the person making the revision.

Sometimes a plant is not sufficiently different from others to be considered a separate species. A major subdivision of a species is described as a subspecies (ssp.), and a lesser division as a variety (var.). In other cases the difference is not even varietal, but only a slight variation in form (forma or f.). Subspecies, varieties, and forms are written after the plant name.

As with other forms of knowledge, botanists are not always in agreement as to how some ferns should be classified. As new information is gathered, more changes will be made in fern classification and names.

The Ferns of Georgia

Georgia is the largest state east of the Mississippi River. To the south and east, the low-lying Coastal Plain contains vast swamps, broad river systems, and low, sandy terrain covering over half the state. Toward the northwest the Piedmont forms a region of dissected, low-lying hills, separated from the Coastal Plains by

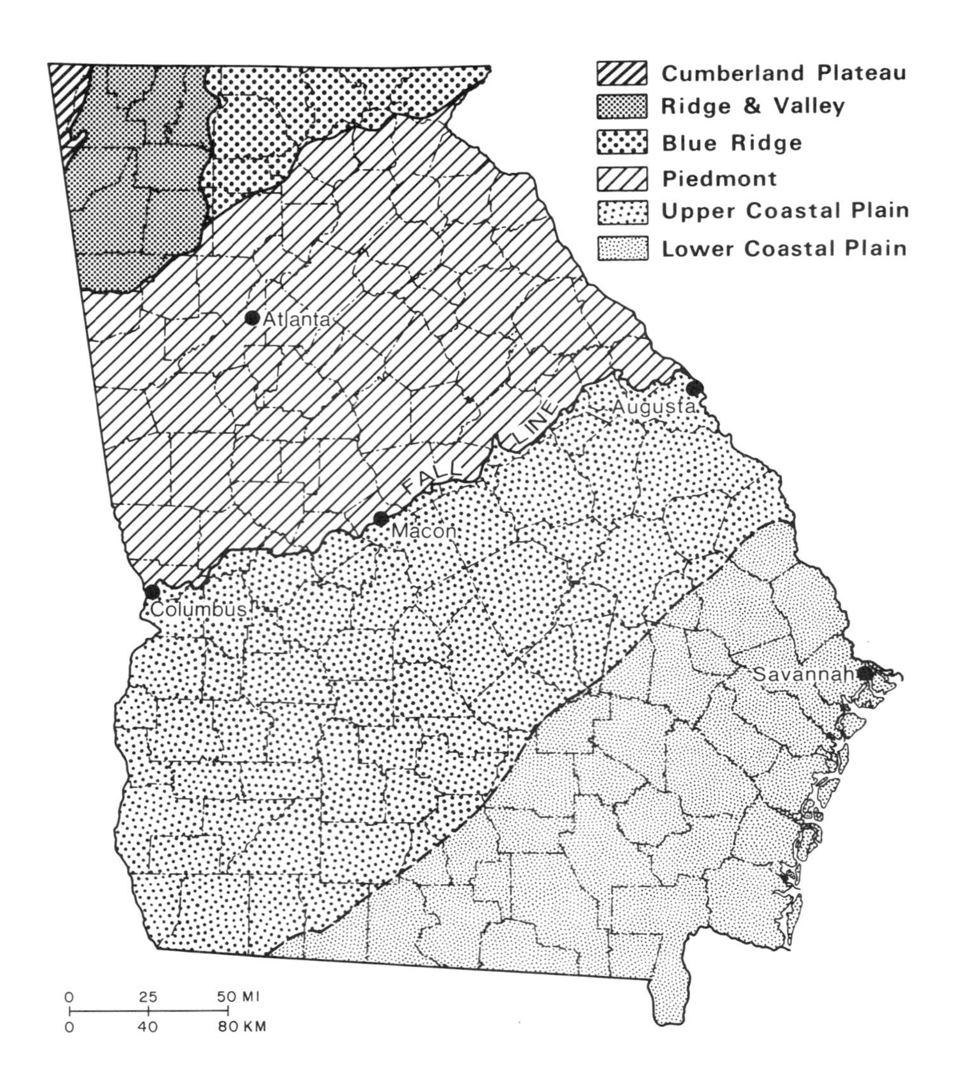
Cumberland Plateau
Ridge & Valley
Blue Ridge
Piedmont
Upper Coastal Plain
Lower Coastal Plain
Atlanta
Augusta
FALL LINE
Macon
Columbus
Savannah
0 25 50 MI
0 40 80 KM

the Fall Line, which runs approximately from Columbus through Macon to Augusta. Because of agricultural practices, the Piedmont is probably the most disturbed of Georgia's physiographic provinces. As the Piedmont rises in elevation along its northern boundary, it borders the Blue Ridge Province in the northeast and the Ridge and Valley Province in the northwest. In the most northwest portion of the state, the Cumberland Plateau is present principally as a part of Lookout Mountain. Elevations in the state range from sea level along the Atlantic coast to nearly 4,800 feet in the Blue Ridge. Climate is mild in southern Georgia with some coastal areas having infrequent frosts, while freezing temperatures and ice and snow are common throughout the winter in the more northern, mountainous regions.

Georgia's pteridophytes reflect the state's physiographic diversity. In the southern and coastal regions are several plants with tropical affinities such as *Blechnum occidentale, Diplazium japonicum, Thelypteris kunthii, T. torresiana, T. hispidula, Lycopodium cernuum,* and *Psilotum nudum.* Conversely, in the northern, mountainous areas are distinctly boreal ferns such as *Dryopteris goldiana, D. intermedia, Osmunda claytoniana,* and *Polypodium virginianum.*

Many persons, both professional and amateur botanists, have contributed to our knowledge of Georgia ferns. The most noteworthy study of Georgia ferns is Rogers McVaugh and Joseph H. Pyron's *Ferns of Georgia.* Published in 1951, this book included the fieldwork of Robert F. Thorne, R. M. Harper, E. W. Graves, Edgar T. Wherry, and others, as well as the authors' extensive travels and fern studies throughout the state in the mid-1930s. This work listed 31 genera, 78 species, and 2 hybrids of pteridophytes in Georgia.

Since the publication of McVaugh and Pyron's book, considerable additional information on the ferns of Georgia has been gathered. Among the many people who have extended our knowledge are Wilbur H. Duncan, Samuel B. Jones, Jr., Wayne R. Faircloth, Juanita N. Faircloth, Nancy C. Coile, Marge White, Harriet Di Gioia, David L. Emory, Alan Cressler, Jocelyn Hill, Stephen M. Bowling, and Marie Mellinger. During the period from 1978 to 1983 Lloyd H. Snyder, Jr., visited all of Georgia's 159 counties and collected over 800 county records.

The number and known ranges of most of Georgia's ferns have been expanded; 5 genera, 40 species, and 10 hybrids have been added. In addition, 8 species listed by McVaugh and Pyron have undergone nomenclatural changes, and specimens for 5 species have been re-identified. In 1980, James G. Bruce, Samuel B. Jones, Jr., and Nancy C. Coile published a current listing of the pteridophytes of Georgia.

The present field guide makes available this new knowledge and provides an up-to-date guide to Georgia ferns and other pteridophytes. Today 36 genera, 119 species, and 12 hybrids have been found in Georgia. As McVaugh and Pyron state in their book: "A work of this kind is never finished. We shall never know

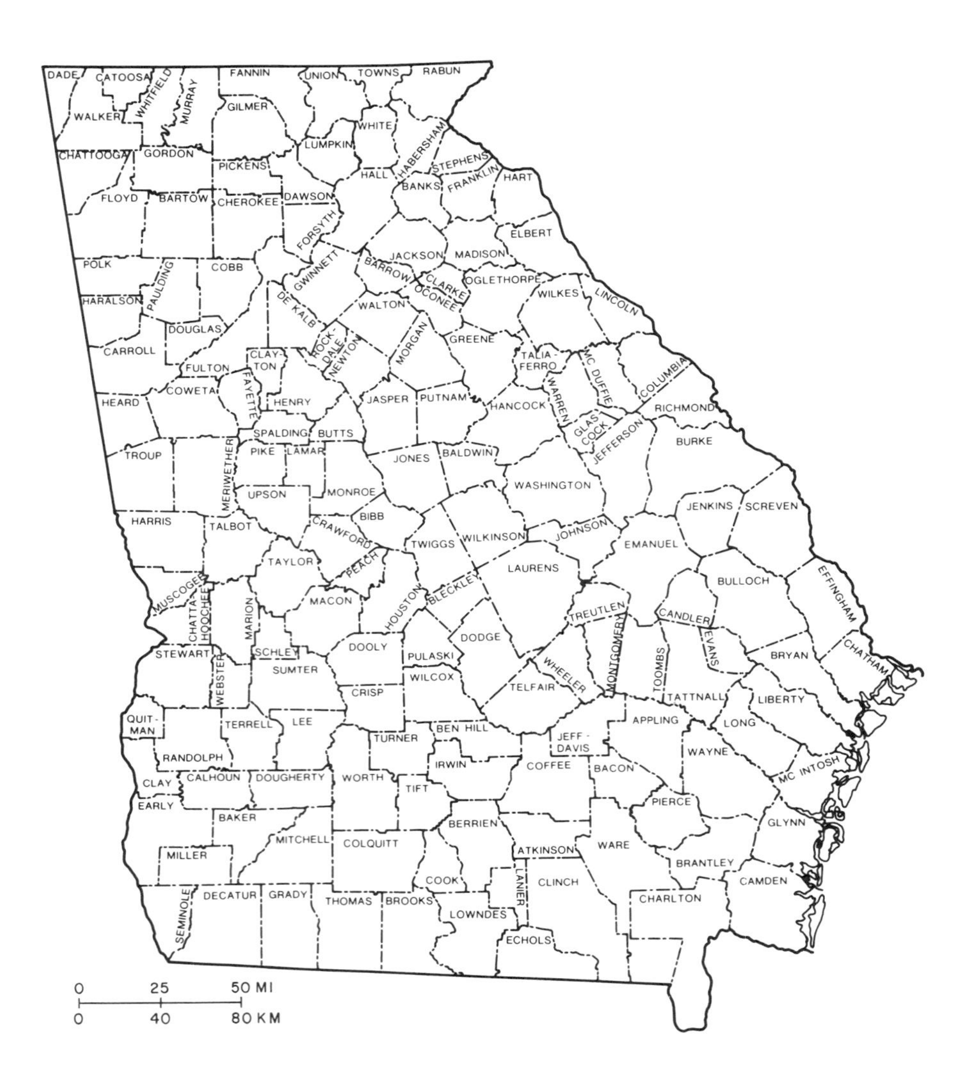
DADE
CATOOSA
WHITFIELD
MURRAY
FANNIN
UNION
TOWNS
RABUN
WALKER
GILMER
WHITE
LUMPKIN
CHATTOOGA
GORDON
HABERSHAM
STEPHENS
PICKENS
HALL
BANKS
FRANKLIN
HART
FLOYD
BARTOW
CHEROKEE
DAWSON
FORSYTH
ELBERT
JACKSON
MADISON
POLK
COBB
PAULDING
GWINNETT
BARROW
CLARKE
OGLETHORPE
HARALSON
DE KALB
OCONEE
WILKES
LINCOLN
WALTON
DOUGLAS
ROCK-DALE
NEWTON
MORGAN
GREENE
CARROLL
CLAY-TON
TALIA-FERRO
MC DUFFIE
FULTON
COLUMBIA
COWETA
FAYETTE
HENRY
JASPER
PUTNAM
HANCOCK
WARREN
RICHMOND
HEARD
SPALDING
BUTTS
GLAS-COCK
JEFFERSON
TROUP
PIKE
LAMAR
JONES
BALDWIN
BURKE
MERIWETHER
UPSON
MONROE
WASHINGTON
JENKINS
SCREVEN
HARRIS
TALBOT
BIBB
CRAWFORD
TWIGGS
WILKINSON
JOHNSON
EMANUEL
TAYLOR
PEACH
LAURENS
BULLOCH
EFFINGHAM
MUSCOGEE
CHATTA-HOOCHEE
MACON
BLECKLEY
MARION
HOUSTON
TREUTLEN
CANDLER
DODGE
DOOLY
SCHLEY
STEWART
PULASKI
MONTGOMERY
EVANS
TOOMBS
BRYAN
CHATHAM
WHEELER
SUMTER
WEBSTER
WILCOX
TELFAIR
CRISP
TATTNALL
LIBERTY
QUIT-MAN
TERRELL
LEE
BEN HILL
JEFF-DAVIS
APPLING
LONG
TURNER
RANDOLPH
IRWIN
COFFEE
BACON
WAYNE
MC INTOSH
CLAY
CALHOUN
DOUGHERTY
WORTH
TIFT
EARLY
PIERCE
BAKER
BERRIEN
GLYNN
MITCHELL
COLQUITT
ATKINSON
WARE
MILLER
BRANTLEY
COOK
CLINCH
CAMDEN
LANIER
SEMINOLE
DECATUR
GRADY
THOMAS
BROOKS
CHARLTON
LOWNDES
ECHOLS
0 25 50 MI
0 40 80 KM

all there is to know about the ferns of Georgia." We hope that our book will stimulate new contributions to the knowledge of ferns. The University of Georgia continues to welcome additional records and information.

Using This Manual

This book has been planned primarily as a field guide to help the amateur learn, identify, and appreciate the ferns of Georgia. In trying to identify ferns, particularly at first, one should avoid young, immature plants, as their characteristics often differ from those of mature plants. The sterile fronds of some ferns (e.g., *Adiantum pedatum, Thelypteris noveboracensis,* and *T. hexagonoptera*) are distinctive enough to make identification easy. In most cases, however, positive identification is often possible only by examining the fertile fronds. Therefore these should be located if at all possible.

Although many ferns can be identified with the naked eye, a small pocket lens of 10 or 20× magnification is often helpful and inexpensive. In a few cases, such as with *Isoetes* and *Azolla,* positive species identification may require laboratory examination with a microscope.

The fern descriptions in this book have been made as brief and yet as complete as possible. Opposite each species' description is a drawing of the fern and a location map. The location maps show counties where ferns have been found. All records are verifiable. Most specimens are on file in the herbarium at the University of Georgia. Some additional records are at the herbaria of Valdosta State College and Georgia Southern College. Literature records of specimens on file at other herbaria have also been included. Although many ferns are found in unusual locations, the normal habitat and location can be helpful in identifying a fern.

As an aid in determining a fern genus, the following chart of quick recognition features of various genera is given. The majority of ferns are arranged first by size, as this is generally the most easily recognized characteristic. It should be remembered, however, that a fern, like other plants, will vary in size and health according to its soil and habitat conditions. For example, although *Osmunda cinnamomea,* grows into a large fern in an ideal situation, it sometimes is much smaller in poor soil. Also, different species in the same genus vary considerably in size, so that this characteristic is often only an approximation. In the following chart, the characteristics given for a particular genus apply only to those species found in Georgia and do not necessarily apply to all species in the genus.

GENERA RECOGNITION FEATURES

Plants mostly small (fronds less than 30 cm long and 5 cm wide), rock-inhabiting:

Blade pinnate, bipinnate, or pinnatifid. Sori elongate along veins on each side of midvein. May become confluent with age. Indusia laterally attached. One X-shaped vascular bundle. **Asplenium,** page 78

Blade entire, long, slender, rooting at tip. Sori scattered along netted veins. **Camptosorus,** page 106

Fronds small to minute. Only one cell thick. Sori marginal in tubular cups. On shaded wet rocks. **Trichomanes,** page 74

Plants small to medium (fronds usually 15 to 45 cm long), mostly rocky areas:

Blade oblong, hairy and/or scaly. Sori under reflexed margins. One vascular bundle in stipe. **Cheilanthes,** page 54

Blade thin-textured, bipinnate-pinnatifid. Stipe pale, glabrous. Sori medial, round. Indusia laterally attached, hoodlike. Two vascular bundles. **Cystopteris,** page 122

Blade bipinnate. Stipe round, hairy, dark. Sori submarginal under inrolled leaf. **Pellaea,** page 62

Blade bipinnate-pinnatifid, oblong to linear. Stipe brown or brown at base, straw-colored above, scaly. Sori medial, round. Indusia cuplike with four lobes. Two vascular bundles. **Woodsia,** page 128

Plants medium (fronds usually 25 to 60 cm long):

Stipe and rachis shiny black. Pinnules fan-shaped or rhomboid, thin-textured. Sori under reflexed margins. **Adiantum,** page 50

Blade pinnate below, pinnatifid above. Sori massed on each side of pinna midvein. **Blechnum,** page 108

Blade pinnate. Pinnae large, entire, leathery. Sori round, prominent, scattered. Indusia peltate. Veins netted. **Cyrtomium,** page 168

Blade pinnate-pinnatifid. Dark green above, lighter below. Sori elongate along veins; at least a few back to back. Indusia prominent, laterally attached. **Diplazium,** page 120

Fronds dimorphic. Sterile pinnatifid. Veins netted. Fertile frond with long narrow segments. Sori in chainlike rows along pinna midvein. **Lorinseria,** page 112

Fronds dimorphic. Sterile leaves pinnatifid. Veins netted. Fertile frond a woody stalk with clusters of beadlike pinnules. **Onoclea,** page 132

Blade pinnate. Fertile frond dimorphic with fertile pinnae smaller and on apical portion of blade. Sori medial, round, prominent. Indusia peltate. Stipe and rachis scaly. **Polystichum,** page 166

Blade pinnate. Pinnae mostly long, narrow. Sori submarginal under reflexed pinnae edges. **Pteris,** page 64

Plants medium to large (fronds usually 30 to 100 cm long):

Blade oval to lance-shaped, pinnate to tripinnate. Sori in rows on each side of midvein. Sometimes hooked over vein. Indusia laterally attached. Two vascular bundles. **Athyrium,** page 114

Blade yellow-green, bipinnate-pinnatifid, hairy. Sori marginal in cuplike indusia. One horseshoe-shaped vascular bundle in cross section. **Dennstaedtia,** page 70

Blade bipinnate to tripinnate, broadly triangular. Pinnae and pinnules vari-shaped. Sori marginal under reflexed margin. Usually dry habitat. **Pteridium,** page 72

Blade firm or leathery, pinnate-pinnatifid to tripinnate. Stipe scaly. Sori round. Indusia kidney-shaped. Five vascular bundles. **Dryopteris,** page 150

Blade pinnate to tripinnate, mostly pinnate-pinnatifid, with tiny, needlelike hairs. Sori medial, round. Indusia kidney-shaped. Two vascular bundles. **Thelypteris,** page 134

Blade pinnate-pinnatifid, leathery. Sori in chainlike rows along areoles on each side of pinnule midvein. Occurs in wet places. **Woodwardia,** page 110

Plants large (fronds usually up to 100 cm or more long):

Fronds clustered. Fronds or pinnae dimorphic. Sporangia in reduced, non-leafy pinnae either at end of fertile stalk or as separate pinnae on fertile blade. One vascular bundle. **Osmunda,** page 34

Fronds long, slender, twining and climbing. Pinnae dimorphic. Sporangia on fingerlike projections of pinnules. **Lygodium,** page 46

Plants epiphytic on trees or growing on rocks:

Blade pinnatifid. Rhizome creeping, mostly on surface. Sori medial, round, no indusia. **Polypodium,** page 40

Fronds entire, narrow, linear, hanging down. **Vittaria,** page 68

Plants with only one or few leaves per plant and fertile spike:

Mostly small, sometimes medium. Sterile blade triangular, pinnate to tripinnate. Fertile stalk taller than sterile, with sporangia in globular clusters. **Botrychium,** page 14

Small. Sterile blade round or elliptic, entire. Taller fertile stalk with double rows of sporangia. **Ophioglossum,** page 24

Plants aquatic:

Floating. Tiny leaves less than 1 mm long, composed of two round lobes. Leaves in two rows along stem. **Azolla,** page 174

Fronds small, resembling grass. No blade or leaflets. **Pilularia,** page 172

Floating. Leaves about 1 cm in diameter, round or oval, in two rows along stem. **Salvinia,** page 178

Other pteridophytes:

Plants small. Forking green stems without leaves or roots. Widely scattered tiny, scalelike bracts. Sporangia solitary, three-chambered. **Psilotum,** page 254

Plants mostly 15 to 30 cm tall with both horizontal and erect stems. Leaves on both stems small, scalelike, with single vein, arranged spirally. Sporangia borne in axils of vegetative leaves or in sporophylls of a cone. **Lycopodium,** page 186

Plants mostly small, creeping, mosslike. Leaves small, scalelike with only central vein. **Selaginella,** page 214

Plants grass- or onion-like tufts. Stem cormlike. Leaves slender with single vein. Habitat, in or by water—often shallow pools on rock outcrops. **Isoetes,** page 234

Aerial stems bamboolike, jointed, grooved, and hollow. Leaves small, scalelike, whorled, fused to form sheath at joints. **Equisetum,** page 250

Ferns

Rattlesnake Fern

Botrychium virginianum (L.) Sw.

Name: In 1753 Linnaeus named this fern *Osmunda virginiana.* Swartz established the genus *Botrychium* in 1801 and transferred this plant to it. *Botrychium* is from a Greek word for "bunch of grapes," alluding to the grapelike clusters of sporangia. Hence also the common genus name of "grapefern." The epithet means "of Virginia." "Rattlesnake fern" is from a fancied resemblance of the sporangial clusters to that reptile's rattles.

Rootstock: Erect. Roots fleshy and numerous.

Fronds: Appear in early spring and last until late fall. Deciduous.

Stipe: Usually 18 to 35 cm long, but very variable. Fleshy.

Blade: 10 to 25 cm long and 15 to 40 cm broad, but very variable in size. Broadly triangular, ternate. Bipinnate to tripinnate. Pinnules deeply incised, pointed, toothed.

Fertile Stalk: 20 to 40 cm long, arising from base of blade. Sporangia densely clustered on branching segments. Appear in late spring or early summer and wither by fall.

Habitat: Moist deciduous woodlands. Usually in rich, well-drained soils, in calcareous and non-calcareous areas.

Range: Common throughout much of North America, Europe, and Asia. Common in northern Georgia, sporadic in the southern part of the state except in the southeastern Coastal Plain.

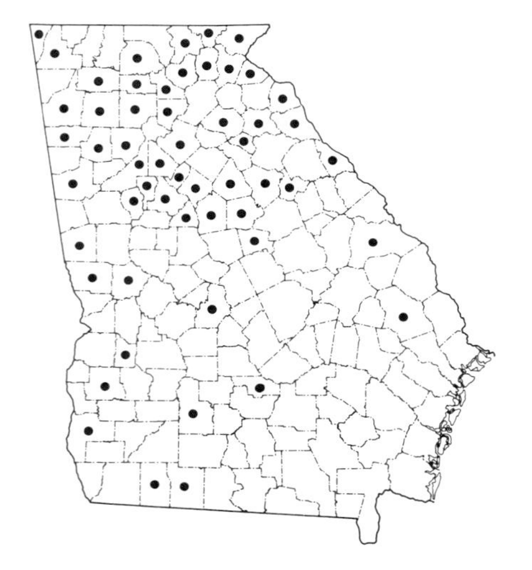

Winter Grapefern

Botrychium lunarioides (Michx.) Sw.

Name: First discovered in South Carolina in 1800, this fern was named *Botrypus lunarioides* by Michaux in 1803. Swartz transferred it in 1806 to *Botrychium.* The epithet means "resembling Lunaris," an old generic name for plants now included in *Botrychium.*

Rootstock: Short, erect, with fleshy, branching roots.

Fronds: Appear in late November. Fronds, fertile spike and all, completely die in March—not to reappear until 7 months later.

Stipe: 2.5 to 5 cm long.

Blade: 2.5 to 5 cm long and 3 to 6 cm broad. Prostrate on ground. Broadly triangular. Bipinnate to tripinnate. Segments small, roundish or fan-shaped, close-set, and finely toothed.

Fertile Stalk: Sterile portion 10 to 15 cm long. Sporangia somewhat wide-spread on branching segments. Develops in spring.

Habitat: Dryish, open woods, old fields, and woody slopes. Soil usually subacidic.

Range: Scattered localities in southeastern United States from North Carolina southward to Florida and westward to Louisiana. In Georgia, in widely separated counties of the Piedmont and the Coastal Plain.

Alabama Grapefern

Botrychium jenmanii Underwood

Name: This fern was discovered in Georgia in 1900. Maxon described it in 1906 from an Alabama specimen, and for many years it was known by his epithet, *alabamense.* In 1983, however, Wagner determined that this was the same species as described by Underwood in 1900, so that his name is the earliest and hence the correct one. The name is given in honor of George Jenman, a British gardener and botanist of Jamaica and British Guiana.

Rootstock: Short, erect, with fleshy, branching roots.

Fronds: Appear in summer. Evergreen.

Stipe: 5 to 15 cm long. Succulent, yellowish to brown.

Blade: 5 to 15 cm long and 8 to 20 cm broad. Broadly triangular, more or less ternate. Three main divisions each twice pinnate at base. Segments small, distant, alternate, somewhat fan-shaped, and finely toothed.

Fertile Stalk: Sterile portion 10 to 30 cm long. Sporangia borne in clusters at ends of numerous, small fertile segments. Developing in late summer, maturing in fall.

Habitat: Dry hillsides under pines, in sandy or red clay soils, and moist situations in wooded ravines.

Range: Limited to North Carolina southward to northern Florida and westward to Louisiana. In Georgia, found in the Piedmont, Blue Ridge, and Upper Coastal Plain. A rather rare fern.

Remarks: *Botrychium jenmanii* is generally believed to be a fertile hybrid of *B. biternatum* and *B. lunarioides,* and it has characteristics of both of these.

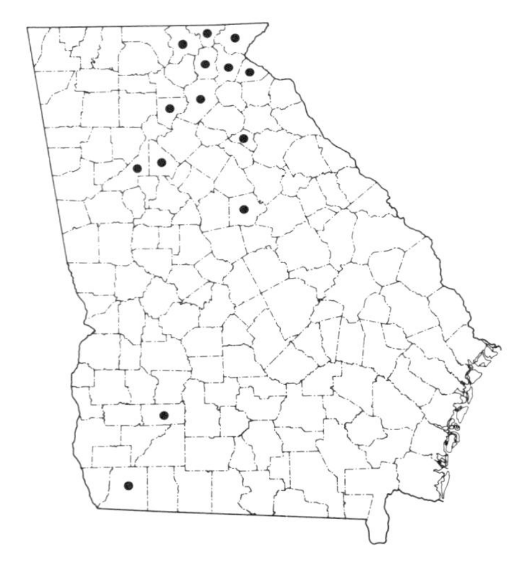

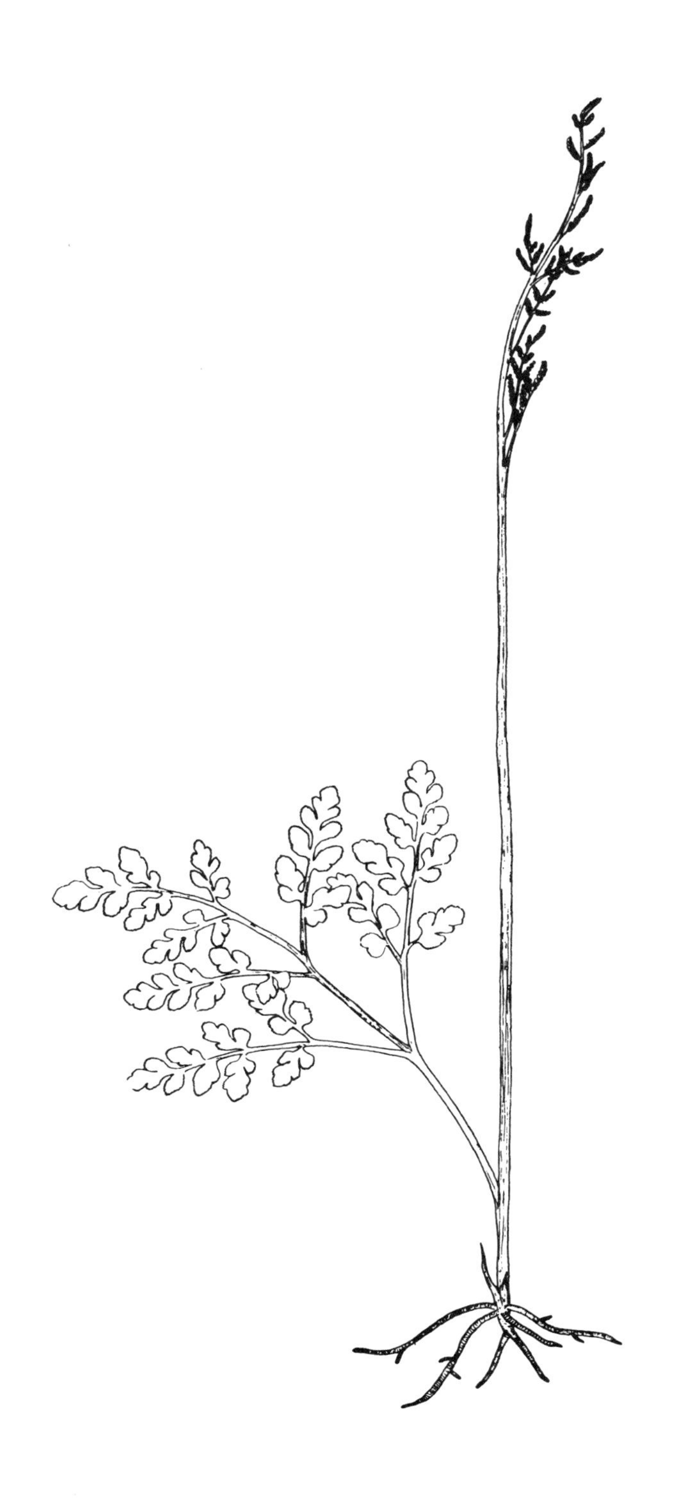

Dissected Grapefern

Botrychium dissectum Spreng.

Name: Sprengel named this fern in 1804 from a Virginia plant. The epithet and common name are appropriate for the much-dissected leaf segments of this fern.

Rootstock: Erect. Roots fleshy and numerous.

Fronds: Appear in summer. Evergreen, bronze in winter and wither in spring.

Stipe: 5 to 20 cm long. Succulent, round, yellow to yellow-green.

Blade: 10 to 40 cm long and 15 to 50 cm broad at base. Broadly triangular, more or less ternate. Bipinnate and tripinnate at base. Segments long, somewhat pointed, several times as long as wide. Margins much-toothed, giving a lacy appearance.

Fertile Stalk: Sterile portion 15 to 45 cm long. Sporangia clustered on branching segments at end of long stalk. Developing in fall and often remaining in winter.

Habitat: Moist soils in woodlands and open situations in roadside thickets and along stream banks.

Range: Common from Canada southward to Georgia and westward to eastern Texas and Missouri. Throughout northern Georgia down through the Piedmont. Very rare in the Coastal Plain.

Remarks: There are two very distinct forms of this species. The more rare f. *dissectum* is finely cut with a lacy appearance. The more common f. *obliquum* is less finely cut and less lacy in appearance. This latter taxon and *Botrychium biternatum* are variable in form and are occasionally difficult to distinguish except by segment shapes and extent of cutting.

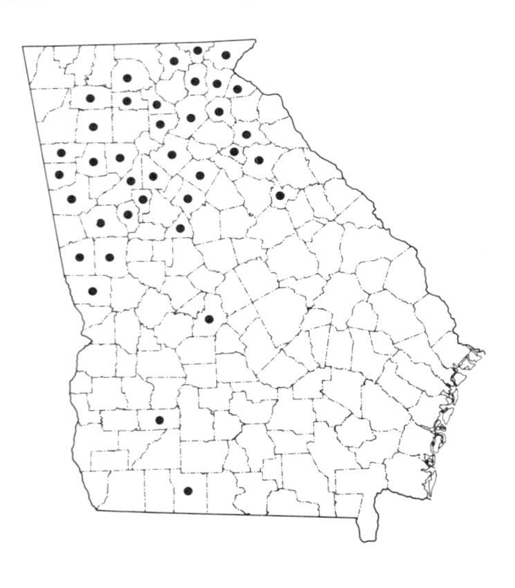

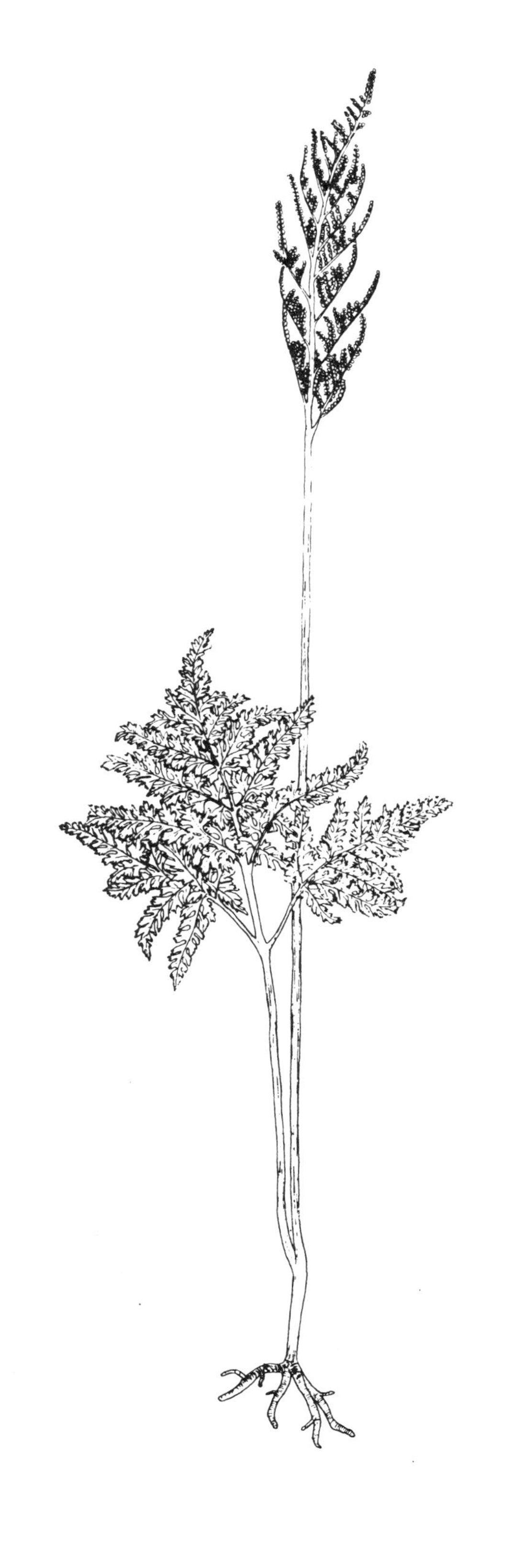

Southern Grapefern

Botrychium biternatum (Sav.) Underw.

Name: In 1707 the French naturalist Savigny named this fern *Osmunda biternata,* and in 1896 Underwood transferred it to *Botrychium.* The epithet means "twice ternate," or "in two groups of threes." It is also commonly called "sparse-lobed grapefern."

Rootstock: Fleshy. Short, vertical.

Fronds: Appear in summer. Evergreen, but edges may turn somewhat bronze in winter.

Stipe: 5 to 15 cm long. Round, succulent. Brown to green.

Blade: 5 to 12 cm long and 8 to 16 cm broad at base. Triangular, ternate (with three, more or less equal divisions). Pinnate to bipinnate. Only two or three pairs of pinnae and basal pinnae with only one or two pairs of pinnules. Pinnules and terminal segments narrow-oblong with pointed tips. Segments variously but sparsely lobed. Margins finely and uniformly toothed.

Fertile Stalk: Sterile portion 20 to 45 cm long. Arising near base of stipe or sometimes below ground level. Sporangia in complex clusters on branching segments at upper end. Withers away in late fall.

Habitat: Mostly in moist woodlands, thickets, and stream banks, but also in drier woods and fields.

Range: Southeastern United States, southward from Maryland and southern Illinois and westward to eastern Texas and Oklahoma. In Georgia, common throughout the Piedmont and frequent in the Blue Ridge and Coastal Plains areas.

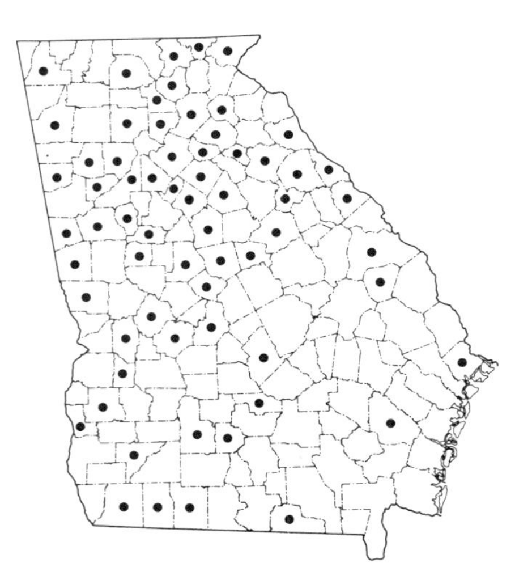

Bulbous Adder's Tongue Fern

Ophioglossum crotalophoroides Walt.

Name: In 1753 Linnaeus applied the ancient name *Ophioglossum,* from the Greek meaning "snake's tongue," to this genus. It comes from the fancied resemblance of the fertile spike to a snake's tongue. Thomas Walter, an English botanist who studied plants in South Carolina, discovered this taxon and named it in 1788. The epithet means "bearing rattles like a snake," referring to the fertile spike. It is also commonly called "dwarf adder's tongue fern."

Rootstock: Globose, bulbous (about the size of a pea, a distinguishing characteristic).

Fronds: 3 to 10 cm long (from ground to tip of fertile spike).

Stipe: (common stalk): 1.5 to 3 cm long.

Fertile Stalk: 1.5 to 7 cm long, with long, bare tip above sporangia.

Blade: Heart-shaped and broadly oval. 1.5 to 3 cm long and 1 to 2 cm wide. One or two blades borne horizontally and nearly flat to the ground.

Sporangia: Beadlike, coherent, in two rows, in spike up to 1 cm long.

Habitat: Common in sandy, grassy areas including open, disturbed habitats such as lawns, roadsides, clearings for utility and pipe lines, and cemeteries. Reaches peak of development in late winter.

Range: Central and South America and southeastern United States from South Carolina to Florida and westward to eastern Texas and Oklahoma. In Georgia, in the Piedmont and the Upper and Lower Coastal Plains.

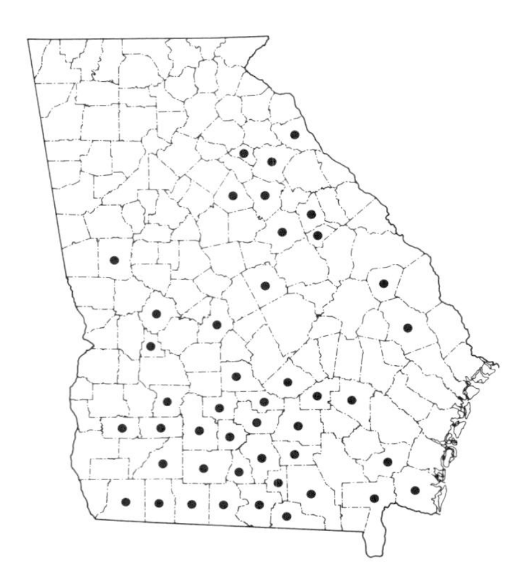

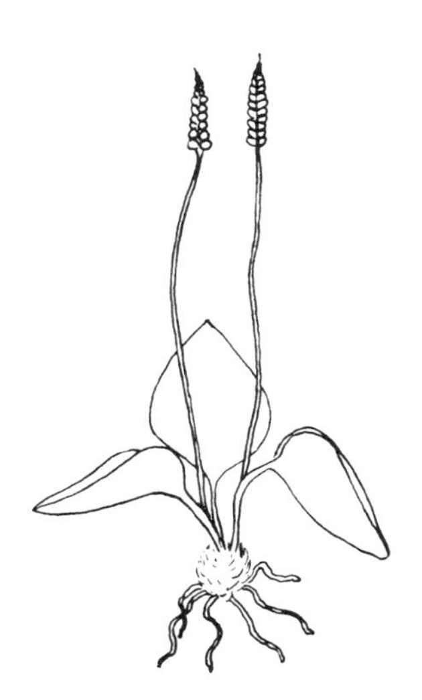

Dwarf Adder's Tongue Fern

Ophioglossum nudicaule L. f. var. *tenerum* Clausen

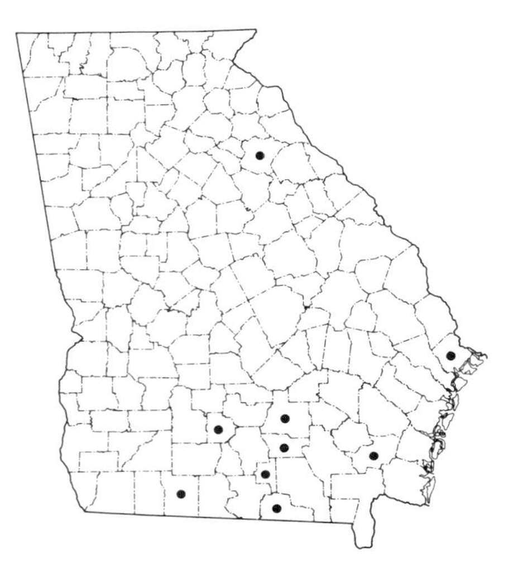

Name: Linnaeus's son named this species in 1781. *Nudicaule* means "bare-stemmed." In 1938 Clausen gave it varietal rank, using a designation applied by Mettenius in 1883. *Tenerum,* meaning "tender," refers to the slender, delicate appearance of the plant. Other common names are "slender" and "least adder's tongue fern."

Rootstock: Fleshy, subglobose.

Fronds: 3.5 to 12 cm long.

Stipe: 0.5 to 1 cm long.

Fertile Stalk: 3 to 11 cm long.

Blade: Oval to oblong. 0.5 to 1.3 cm long and 0.3 to 0.6 cm broad. With an acute tip and a wedge-shaped base. Extra blade occasionally present.

Sporangia: Beadlike, coherent in spike up to 1 cm long.

Habitat: Moist habitats of sandy and peaty soils including grassy areas, seepages, and sandy areas in cemeteries. Frequent during spring rains. Often in colonies.

Range: Tropical areas of America, Africa, Asia, and Australia. In the United States from South Carolina across the southern Gulf states to eastern Texas. In Georgia, primarily in the Lower Coastal Plain.

Limestone Adder's Tongue Fern

Ophioglossum engelmannii Prantl

Name: This fern was recognized as a distinct species by George Engelmann of St. Louis and was named in his honor by the German botanist Karl Prantl in 1883. Also commonly called "Engelmann's adder's tongue fern."

Rootstock: Fleshy, cylindrical.

Fronds: 5 to 20 cm long.

Stipe: 3 to 8 cm long.

Fertile Stalk: 2 to 12 cm long.

Blade: Oblong. 3 to 10 cm long and 1.5 to 3 cm broad. Slightly pointed at both ends, tipped by short but conspicuous, bristlelike projection. Veins form two sets of areoles—fine ones bounded by coarse ones.

Sporangia: Beadlike, coherent, in 1 to 2.5 cm long spike.

Habitat: In colonies in limestone areas. Wet depressions and limestone sinks, pockets of gravel or clay or slopes with sparse vegetation over calcareous shale or limestone outcrops.

Range: Mexico and southern United States from Virginia and northern Florida westward to Missouri and Arizona. In Georgia, abundant in the "Cedar Barrens" of Walker and Catoosa counties.

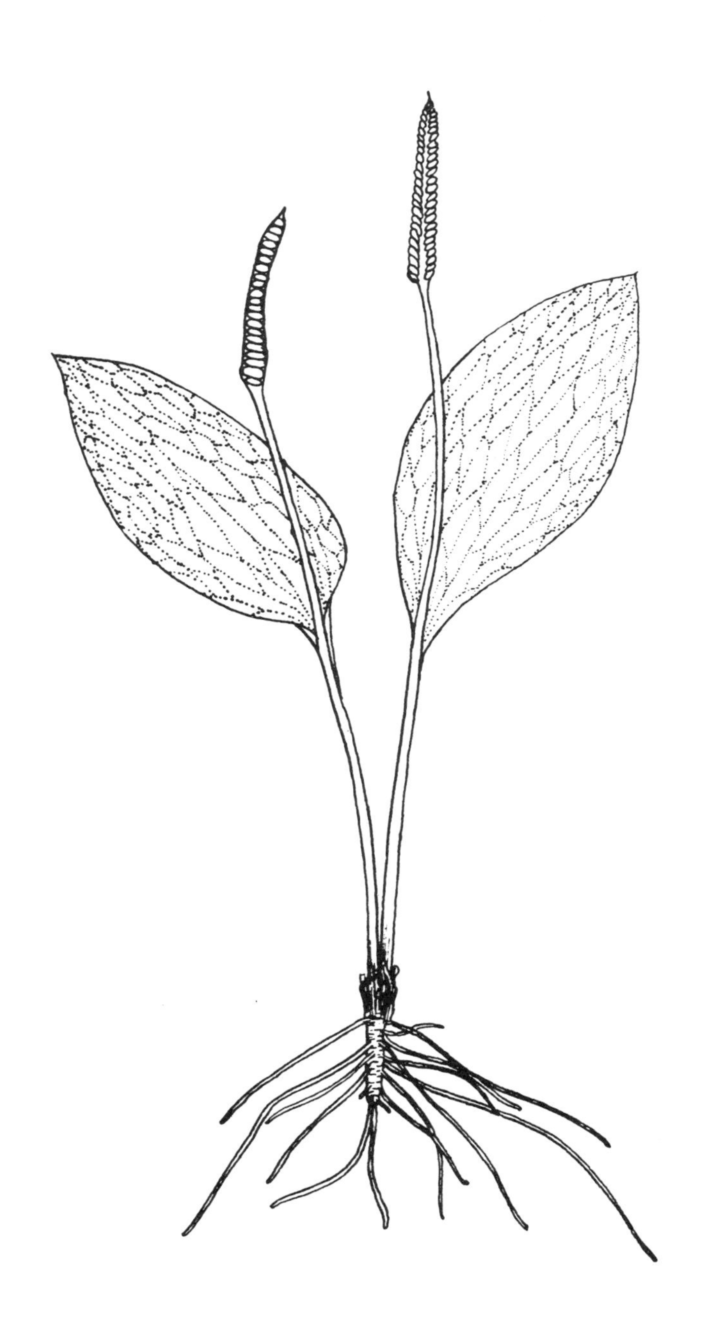

Stalked Adder's Tongue Fern

Ophioglossum petiolatum Hooker

Name: Sir William Hooker, a British taxonomist, named this fern in 1823. *Petiolatum* means "with a stalk at the base of the blade."

Rootstock: Cylindrical with small roots, some bearing adventitious buds.

Fronds: 5 to 20 cm long.

Stipe: 3 to 11 cm long.

Fertile Stalk: 2 to 9 cm long.

Blade: Lance-shaped, abruptly expanding from a short stalk and tapering upward. 2.5 to 7.5 cm long and 1.5 to 3 cm broad.

Sporangia: Beadlike, coherent in 1 to 4 cm long spike.

Habitat: Grassy openings in pinelands, moist meadows, and semishaded alluvial soils; under highway bridges.

Range: Tropical to mild temperate regions. Southeastern United States from South Carolina southward to Florida and westward to eastern Texas and Arkansas. In Georgia, in almost all counties of the Lower Coastal Plain.

Remarks: Prior to 1966 the only record of *Ophioglossum* from the Georgia Coastal Plain was *O. nudicaule* from Effingham County. In that year Juanita Norsworthy, following an intensive study of the pteridophytes of Echols County, reported finding there both *O. nudicaule* and *O. petiolatum.* Since then, she and her husband, Wayne R. Faircloth, and others have made extensive searches for *Ophioglossum* in the Coastal Plain. As a result of these studies, *O. crotalophoroides* has been found in 44 counties, *O. nudicaule* in 9 counties, and *O. petiolatum* in 26 counties. In addition, during their search for *Ophioglossum, Botrychium lunarioides* was found in two south-central Georgia counties. There is reason to believe that additional sites of these ferns exist. It is best to look for them in late winter to early summer before they have been covered up by larger plants. It is also frequently necessary to look for them at ground level, as they are rather difficult to see.

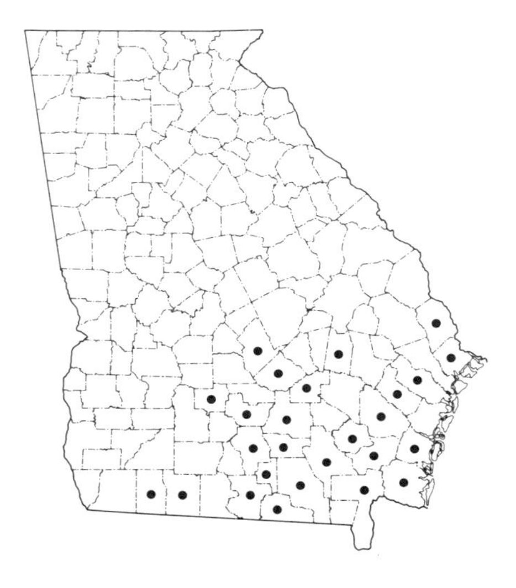

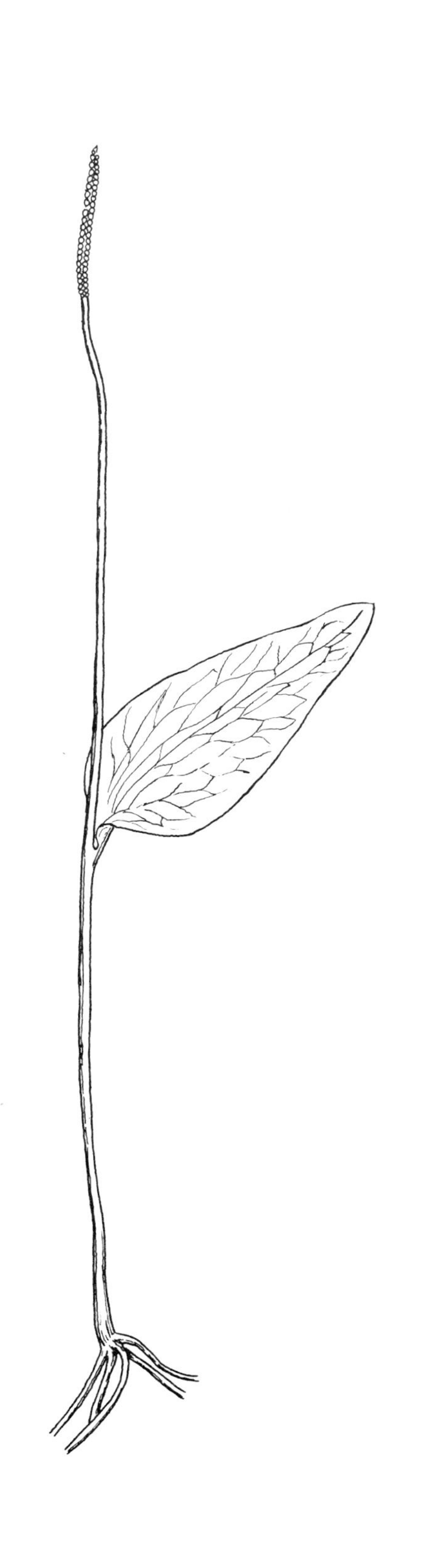

Southern Adder's Tongue Fern

Ophioglossum pycnostichum (Fern.) Löve and Löve

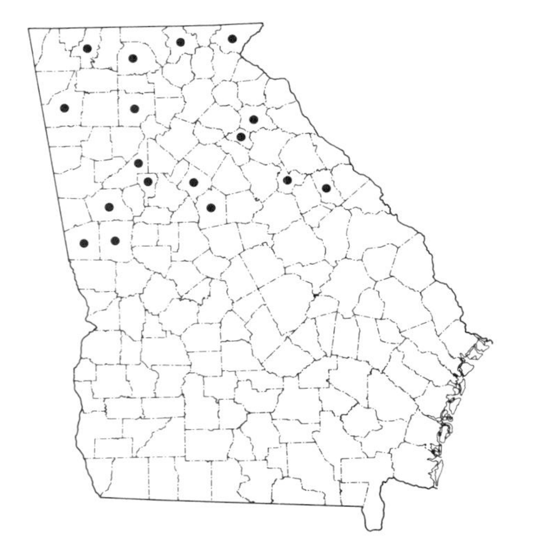

Name: In 1753 Linnaeus named *Ophioglossum vulgatum* ("common"). In 1939 Fernald described the southern variety *O. vulgatum* var. *pycnostichum,* and Askell and Doris Löve raised it to species rank in 1977. *Pycnostichum* means "with crowded rows (of sporangia)."

Rootstock: Fleshy, cylindrical.

Fronds: 8 to 30 cm long.

Stipe: 5 to 12 cm long.

Fertile Stalk: 3 to 18 cm long.

Blade: Oval. 3 to 8 cm long and 2 to 4 cm broad. Rounded at top with broad, abruptly tapering base.

Sporangia: Beadlike, coherent in spike 1 to 2 cm long.

Habitat: Usually in floodplains along streams and in swamps, and also in partially shaded meadows and grassy thickets.

Range: The southern adder's tongue fern is found primarily in the southern and western United States. It occurs throughout the northern section of Georgia above the Fall Line.

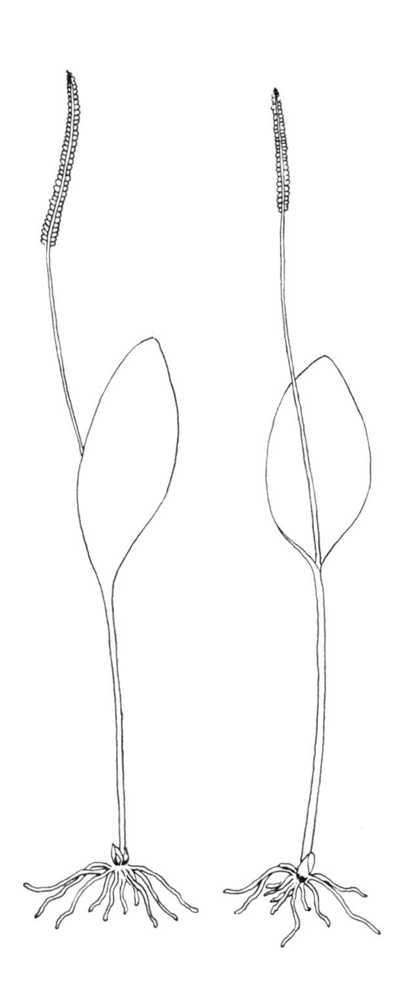

Cinnamon Fern

Osmunda cinnamomea L.

Name: The generic name, selected by Linnaeus, was probably taken from the name of the Saxon god Osmunder. In 1753 Linnaeus gave the specific epithet to Maryland specimens, in reference to the large, cinnamon-colored fertile fronds.

Rootstock: Stout, short-creeping, erect. Roots matted, wiry.

Fronds: 60 to 120 cm long. Clustered, deciduous, dimorphic. Fertile fronds produced in spring (or sometimes in fall as well), without any green, leaflike blade, soon withering.

Stipe: 25 to 45 cm long. Green or straw-colored. Fuzzy, round, grooved in front.

Rachis: Smooth, green. Woolly, especially early in season.

Blade: Sterile, 35 to 75 cm long and 12 to 25 cm wide. Lanceolate with acuminate apex. Widest about one-third of the way up from base. Pinnate-pinnatifid.

Pinnae: Alternate, lanceolate, deeply lobed. Distinctive tuft of rusty hairs on lower side at base of each pinna. Pinnules obliquely acute.

Fertile Fronds: Erect. Apex having strongly ascending pinnae and ascending pinnules, that bear masses of brown naked sporangia, which are green at first, but soon turn brown, shed their spores, and wither away.

Habitat: In wet, somewhat acidic soil. In swamps, sphagnum bogs, wet woods and rocks, and along stream banks.

Range: Common throughout eastern temperate North America. In Georgia, conspicuous in swampy woodlands and found in most counties.

Remarks: Osmunda fiber, which is the old roots and stipe tissue, is commonly dug and used as a potting medium for orchids and other epiphytes.

O. cinnamomea forma *frondosa* Britton is an occasional intermediate between fertile and sterile fronds, as the fronds on this fern have both fertile and sterile pinnae. Injury to the plant from environmental trauma may explain the occurrence of both fertile and sterile pinnae. Specimens of *O. cinnamomea forma frondosa* were collected by Wayne Faircloth in Brantley County in 1982.

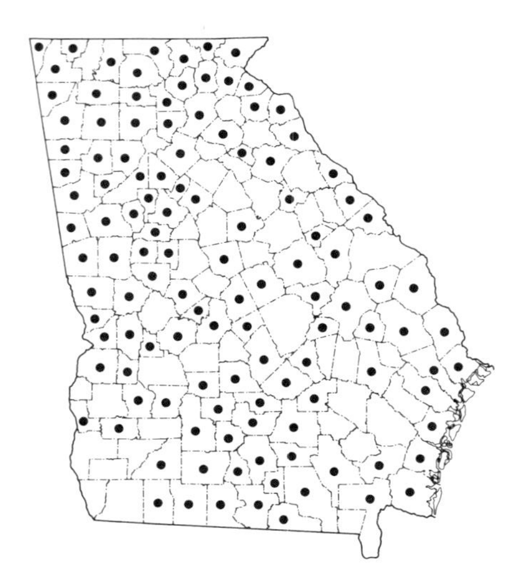

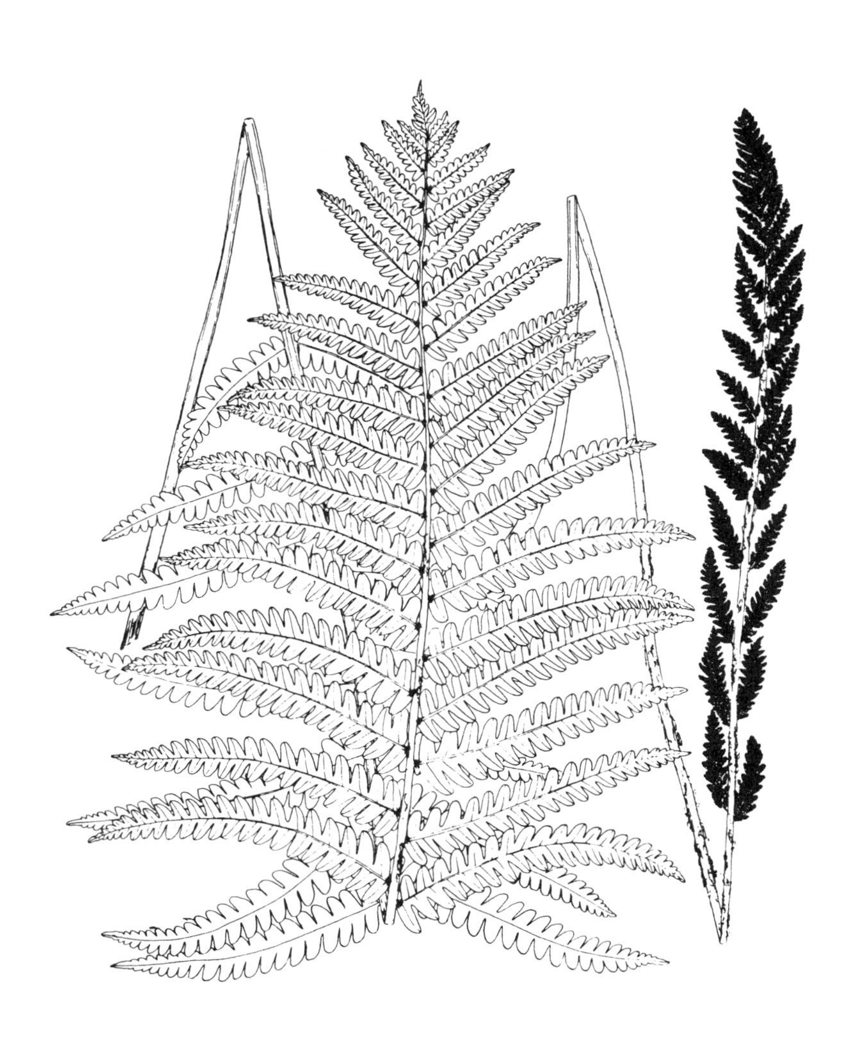

Interrupted Fern

Osmunda claytoniana L.

Name: In 1753 Linnaeus named this fern in honor of the botanist John Clayton, who collected the species in Virginia. The common name refers to the way the group of fertile pinnae interrupt the regular rows of sterile ones.

Rootstock: Large, erect, stout.

Fronds: 60 to 120 cm long. Clustered. Fertile fronds characterized by sterile pinnae interspaced by fertile pinnae.

Stipe: 20 to 40 cm long. Smooth, green or yellow, grooved.

Rachis: Smooth, green or yellow, grooved.

Blade: 40 to 80 cm long and 15 to 30 cm broad. Lanceolate-ovate to elliptic with abrupt apex. Pinnate-pinnatifid. Sterile blade shorter. Fertile blade with both sterile and fertile pinnae. Fertile blade has about 3 pairs of sterile pinnae at base, 1 to 6 pairs of fertile pinnae, and 16 to 19 pairs of sterile pinnae above.

Sterile Pinnae: Oblong, alternate, ascending. Deeply cut into obtuse, blunt-tipped, separated or semi-overlapping pinnules.

Fertile Pinnae: Leafy tissue replaced by short clusters of sporangia resembling miniature grapes. Green at first, turning dark brown as spores are released and then withering away.

Habitat: Rich, wooded slopes. Moist, but not necessarily wet situations in circumneutral or slightly acidic soil. Grows well in fairly open areas.

Range: Widespread in northeastern North America, ranging southward to the mountains of northern Georgia. In Georgia, mostly confined to rich mountain woods at fairly high elevations in the northern counties.

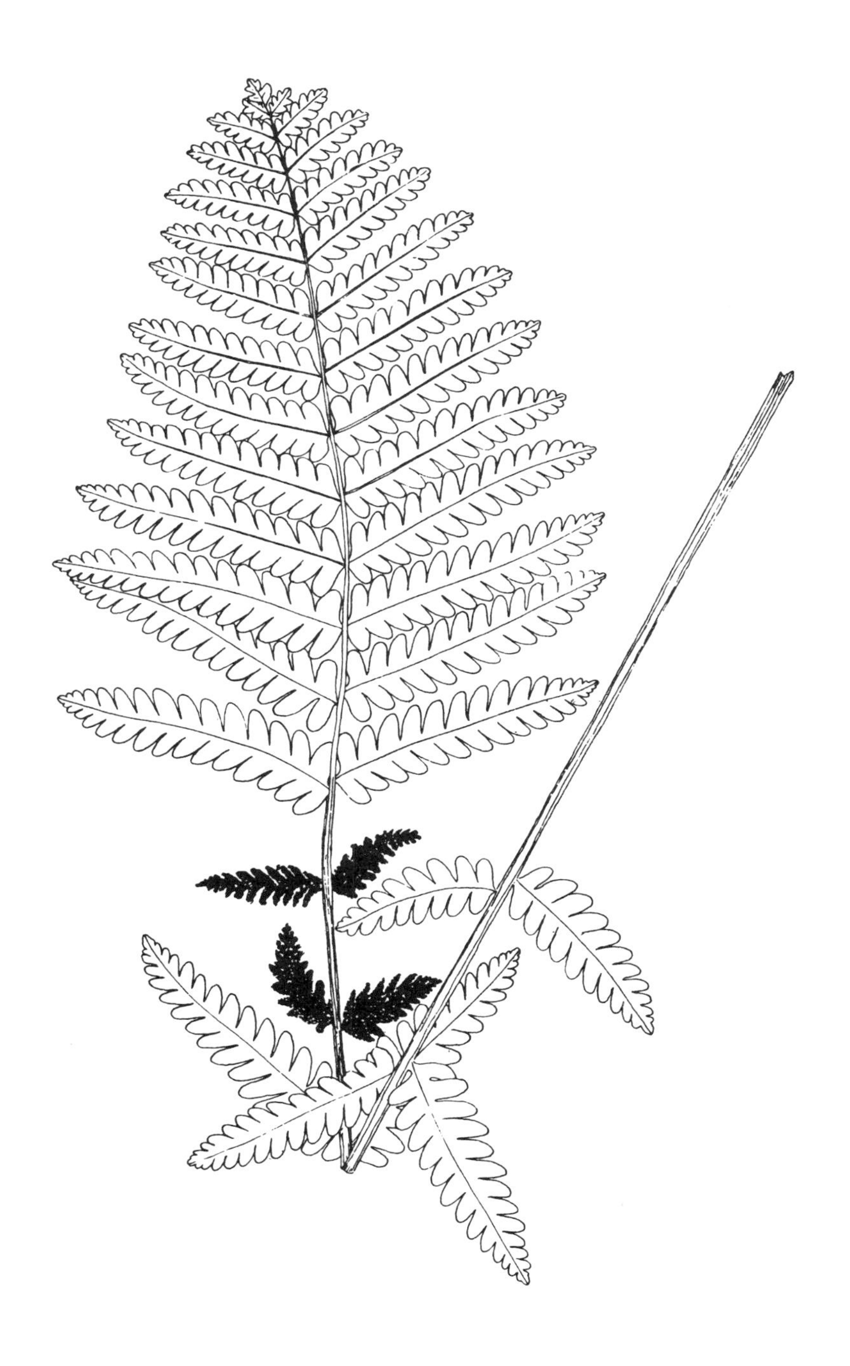

Royal Fern

Osmunda regalis L.

Name: Linnaeus named this in 1753, his epithet *regalis,* or "royal," referring to the handsome fronds. As the American plant is not identical to the European one, in 1856 Asa Gray proposed to call the American variant *spectabilis,* meaning "showy," a designation that is sometimes used.

Rootstock: Stout, short-creeping, erect.

Fronds: 60 to 150 cm long. Clustered. More nearly resemble the leaves of a locust tree than a fern.

Stipe: 30 to 75 cm long. Smooth, straw-colored.

Rachis: Green or straw-colored with few scattered hairs.

Blade: 30 to 75 cm long and 25 to 50 cm wide. Bipinnate.

Sterile Pinnae: Wide-spaced, almost opposite, oblong, ascending. Pinnules wide-spaced, alternate, oblong with oblique base and obtuse tip.

Fertile Pinnae: Lack leafy tissue, restricted to frond tip. Sporangia in naked, short-stalked clusters. Green at first, later brown.

Habitat: In wet woods, cypress and creek swamps, springy depressions and along rocky streams. Soil usually acidic.

Range: Both northern and southern hemispheres of the Old and New worlds. Widespread in eastern North America. Frequent throughout nearly all of Georgia.

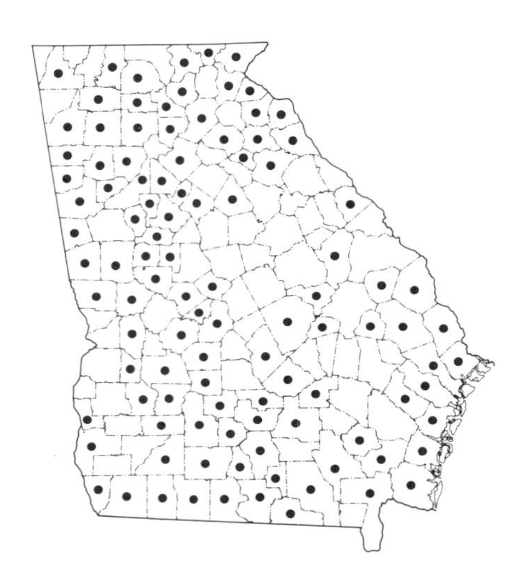

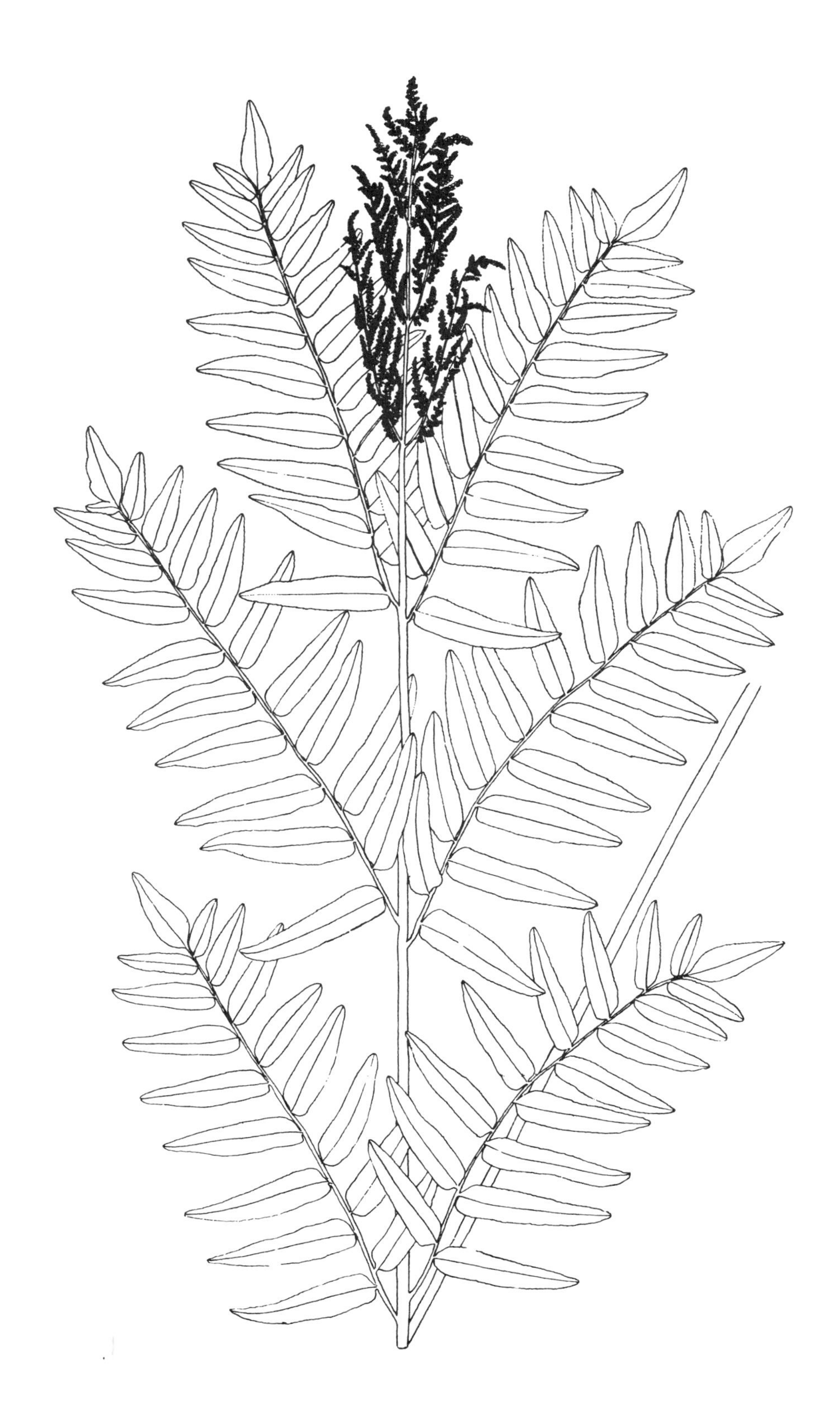

Rockcap Fern

Polypodium virginianum L.

Name: In 1753 Linnaeus established the genus *Polypodium,* from the Greek for "many feet," referring to the footlike branching of the rootstock. He described this fern, as well as the European *Polypodium vulgare,* in his *Species Plantarum* in 1753. While he distinguished between the European and American plants, and most modern botanists concur, some botanists treat them as merely two varieties of the same species. *Virginianum* means "of Virginia." This fern is also called "common polypody."

Rootstock: Creeping, usually partially exposed. Brown scales.

Fronds: 10 to 35 cm long. Produced singly along rootstock. Evergreen.

Stipe: 4 to 15 cm long. Smooth, green.

Rachis: Smooth, green. Flat above.

Blade: 6 to 20 cm long and 2 to 6 cm broad. Thick textured. Bright green above, light green below. Smooth on both sides. Pinnatifid.

Pinnae: 11 to 18 pairs. Oblong, with tip narrowly rounded to pointed. Winged at axis.

Sori: Round, large, and conspicuous. No indusia. In rows on each side of pinna midvein. Nearer margin than midvein. White to yellowish brown when young, darken with age.

Habitat: Mostly on rocks, usually not exposed ones, but partially sheltered boulders and ledges.

Range: Eastern and central North America. In Georgia, common in northern counties. Southernmost, known locations are a rocky bluff in Clarke County, a rocky outcrop along Sope Creek in Cobb County, and in Carroll County.

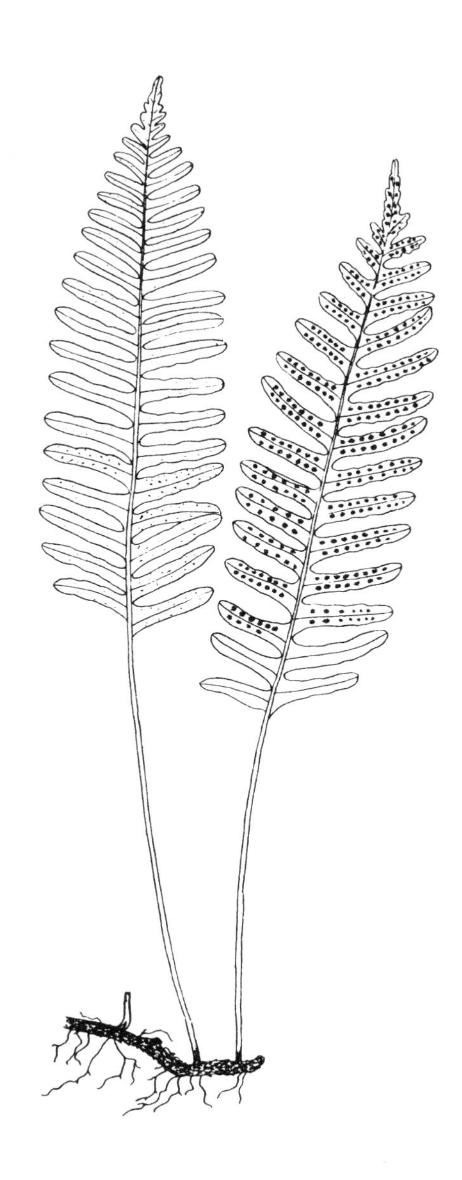

Resurrection Fern

Polypodium polypodioides (L.) Watt

Name: Linnaeus named *Acrostichum polypodioides* in 1753, the epithet meaning "like a *polypodium*." Watt in 1867 transferred it to the *Polypodium* genus, so that the scientific name means "a Polypodium-like Polypodium." The common name of "resurrection fern" is given because the fronds curl up in dry weather and appear to be dead, and are at once revived by rain.

Rootstock: Slender, long-creeping, often exposed.

Fronds: 5 to 18 cm long. Produced singly, distant. Evergreen. Curl up when dry, revive with rain.

Stipe: 2 to 8 cm long. Densely scaled.

Rachis: Green, flat above. Scaly below.

Blade: 3 to 10 cm long and 1.5 to 3.5 cm wide. Oblong. Dark green and smooth above. Densely scaled and silvery brown below. Deeply pinnatifid.

Pinnae: 7 to 14 pairs. Oblong with obtuse tips.

Sori: Round, no indusia. Near margins of pinnae.

Habitat: Epiphytic on the trunks and upper branches of oaks and other rough-barked trees. Forms dense mats on rocks, old shingled roofs, and rotten wood.

Range: Abundant in the southeastern United States and in tropical America. In Georgia, one of the most common ferns and found throughout the state. Especially prolific on trees in the Atlantic coastal areas.

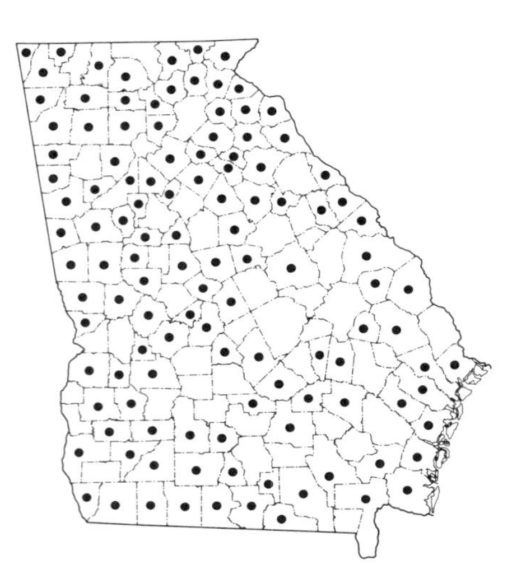

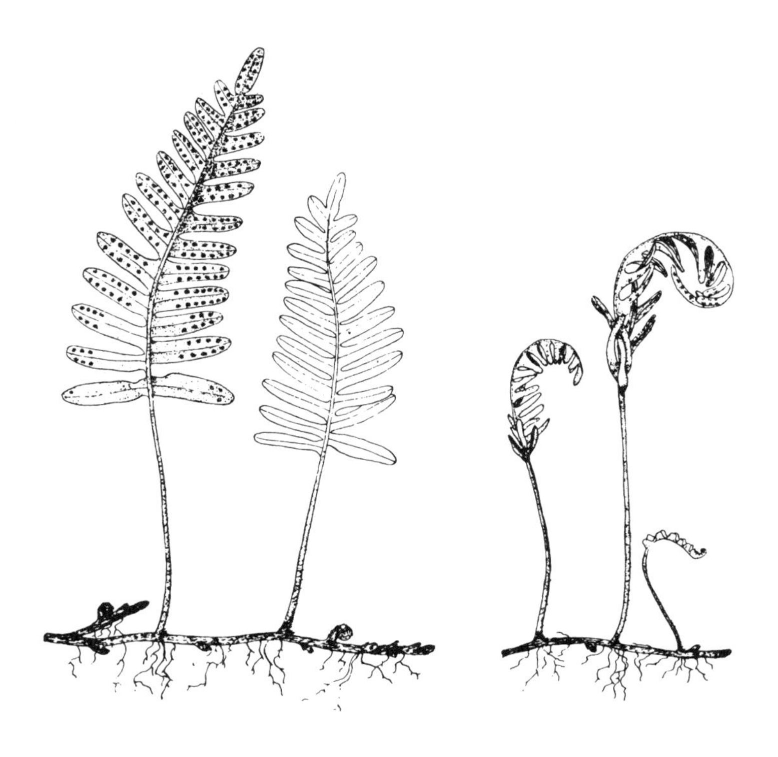

Goldfoot Fern

Polypodium aureum L.

Name: In 1753 Linnaeus named this species in reference to the "golden" appearance of the rhizome. It is also commonly called "golden polypody."

Rootstock: Epiphytic. Stout, creeping on soil surface. 1.5 to 3 cm thick, covered with hairy, reddish orange scales giving the appearance of an animal's furry foot.

Fronds: 40 to 90 cm long. Widely spaced, arching.

Stipe: 20 to 40 cm long. Round, smooth, slightly grooved. Dark, shining brown.

Rachis: Flat on top. Green or brown.

Blade: 20 to 50 cm long and 20 to 40 cm wide. Bluish, becoming yellow-green. Pinnatifid.

Pinnae: 4 to 9 pairs and a terminal one up to 30 cm long. Long, oblong with pointed tips. Winged at rachis. 3 to 5 cm wide. Veins netted with two veins feeding each sorus.

Sori: Round, prominent, wide-spaced in one or two rows on each side of pinna midvein.

Habitat: Epiphytic on palmettos, oaks, and crumbly limestone. Commonly grown as a houseplant, especially in hanging baskets.

Range: Common in tropical America and Florida. In Georgia, has escaped from cultivation in a few coastal counties.

Climbing Fern

Lygodium palmatum (Bernh.) Sw.

Name: The Swedish taxonomist Olof P. Swartz established *Lygodium* in 1801. The genus name comes from a Greek word meaning "flexible," in allusion to the twining leaf-rachis. In 1801 Bernhardi named *Gisopteris palmatum,* and Swartz transferred it in 1806 to *Lygodium.* The epithet is appropriate, as the spreading divisions of the sterile pinnules suggest the fingers of the hand. Commonly this plant is also called "American climbing fern" and "Hartford fern." To conserve a colony near that city, the Connecticut legislature in 1869 passed the first conservation law to protect a fern.

Rootstock: Slender, widely creeping underground, black, wiry, and hairy.

Fronds: 60 to 120 cm long. Arise singly from rootstock. Trailing or twining on other vegetation.

Stipe: 5 to 15 cm long. Wiry. Brownish or greenish with dark base.

Rachis: Straw-colored or green.

Blade: 7 to 15 cm wide. Bipinnate.

Pinnae: Dimorphic. Alternate, widely separated. Sterile pinnae are borne on a forked stalk that branches off rachis, are equally divided into paired palmate pinnules that are 2 to 6 cm broad, with 4 to 6 long elliptical lobes and 2 short basal lobes. Fertile pinnae are on the upper part of the frond, the halves divided several times, and the segments smaller than the sterile ones.

Sori: Sporangia borne on fingerlike ultimate segments of fertile pinnae.

Habitat: Moist woodlands and thickets, wet slopes, sandy bogs and swamps. Soil rich in humus and strongly acidic.

Range: A rare and local fern confined to the eastern and southern United States. In Georgia, only in four northern counties.

Remarks: William Bartram, a gifted Quaker naturalist, traveled in Georgia in 1773. His famous *Travels* contains a reference to what must be the climbing fern: ". . . *filix scandens,* perhaps a species of *trichomanes;* the leaves are palmated, or radiated; it climbs and roves above, on shrubs, in moist ground." It is not clear just where Bartram saw the plant, but it would seem to have been between the Broad and Oconee rivers, not far from the present site of Athens.

Japanese Climbing Fern

Lygodium japonicum (Thunb.) Sw.

Name: Thunberg discovered this taxon in Japan and in 1784 named it *Ophioglossum japonicum.* In 1801 Swartz transferred it to *Lygodium.* The epithet, of course, means "of Japan.".

Rootstock: Slender, widely creeping, black and wiry.

Fronds: 60 to 600 cm long. Arise singly. Trailing and twining.

Stipe: 2 to 10 cm long. Wiry, green or straw-colored.

Rachis: Long, twining, green or straw-colored.

Blade: 16 to 30 cm wide. Pinnately divided into distinct segments.

Pinnae: Dimorphic. Triangular, 8 to 15 cm long and 5 to 8 cm broad at base. Distant. Steril pinnae on lower part of blade, forking into broad, paired segments that are long-triangular and pinnately divided into several also long-triangular segments. Fertile pinnae on upper part of blade are similar but with more and usually smaller segments, which have fingerlike projections around the margins on which sporangia are borne.

Sori: Sporangia in double rows on marginal, fingerlike lobes of fertile pinnules.

Habitat: Mostly disturbed, circumneutral soil of thickets, open woods, roadsides, banks of ditches and streams, and under highway bridges.

Range: A native of Asia that has escaped to the southern states from North Carolina southward to Florida and westward to Texas. In Georgia, common throughout the Coastal Plain, especially in the southwest.

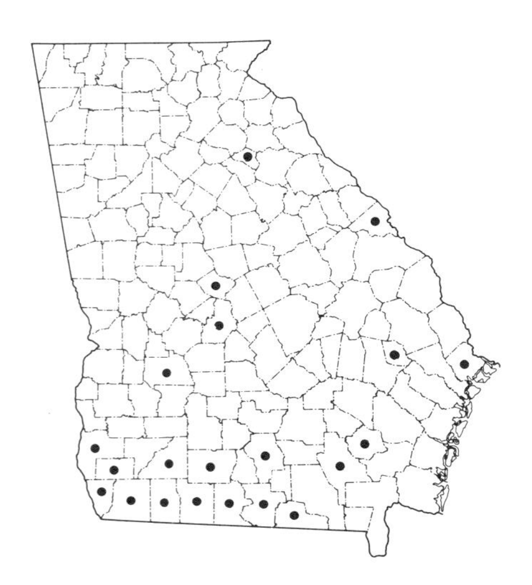

Northern Maidenhair Fern

Adiantum pedatum L.

Name: *Adiantum*, meaning "not to wet," was the Greek name for a fern whose leaves shed water. Linnaeus adopted the name in 1753, when he also named the present species from American material. The epithet, meaning "decreasing outward, like the toes," refers to the outline of the leaf blade.

Rootstock: Stout, short-creeping. Frequently with old stipe bases.

Fronds: 35 to 65 cm long. Erect stalk and more or less horizontal blade. Deciduous, delicate, killed by first frost.

Stipe: 20 to 40 cm long. Shiny, dark brown to black. Smooth except for a few scales at base.

Rachis: Smooth, shiny dark brown. Stipe forks to form two subequal curving rachises.

Blade: 15 to 25 cm long and 20 to 40 cm wide. Semicircular or fan-shaped. Flat, horizontal to ground. Each rachis bears 5 to 7 progressively smaller pinnae along one side.

Pinnae: 5 to 20 cm long. Long, oblong. Pinnules alternate, mostly oblong, 1 to 2 cm long, with smallest ones triangular.

Sori: Oblong. On underside of reflexed margin.

Habitat: Rich, moist, woodland slopes with good drainage, shade, and ample humus. Prefers some lime.

Range: Common throughout Canada and the northern half of the United States. Common in the mountains of northern Georgia, and found in a few scattered Coastal Plain counties.

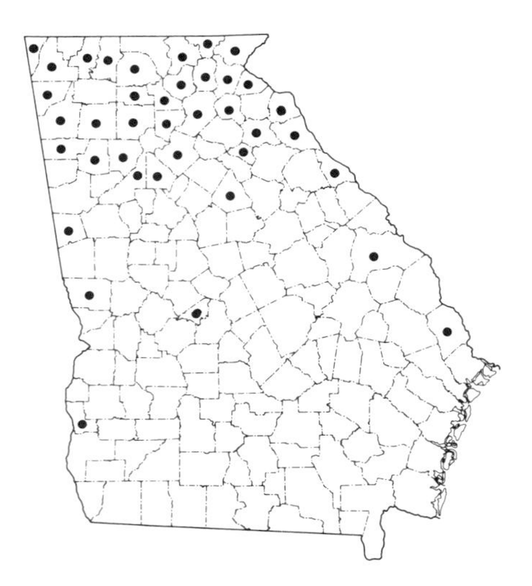

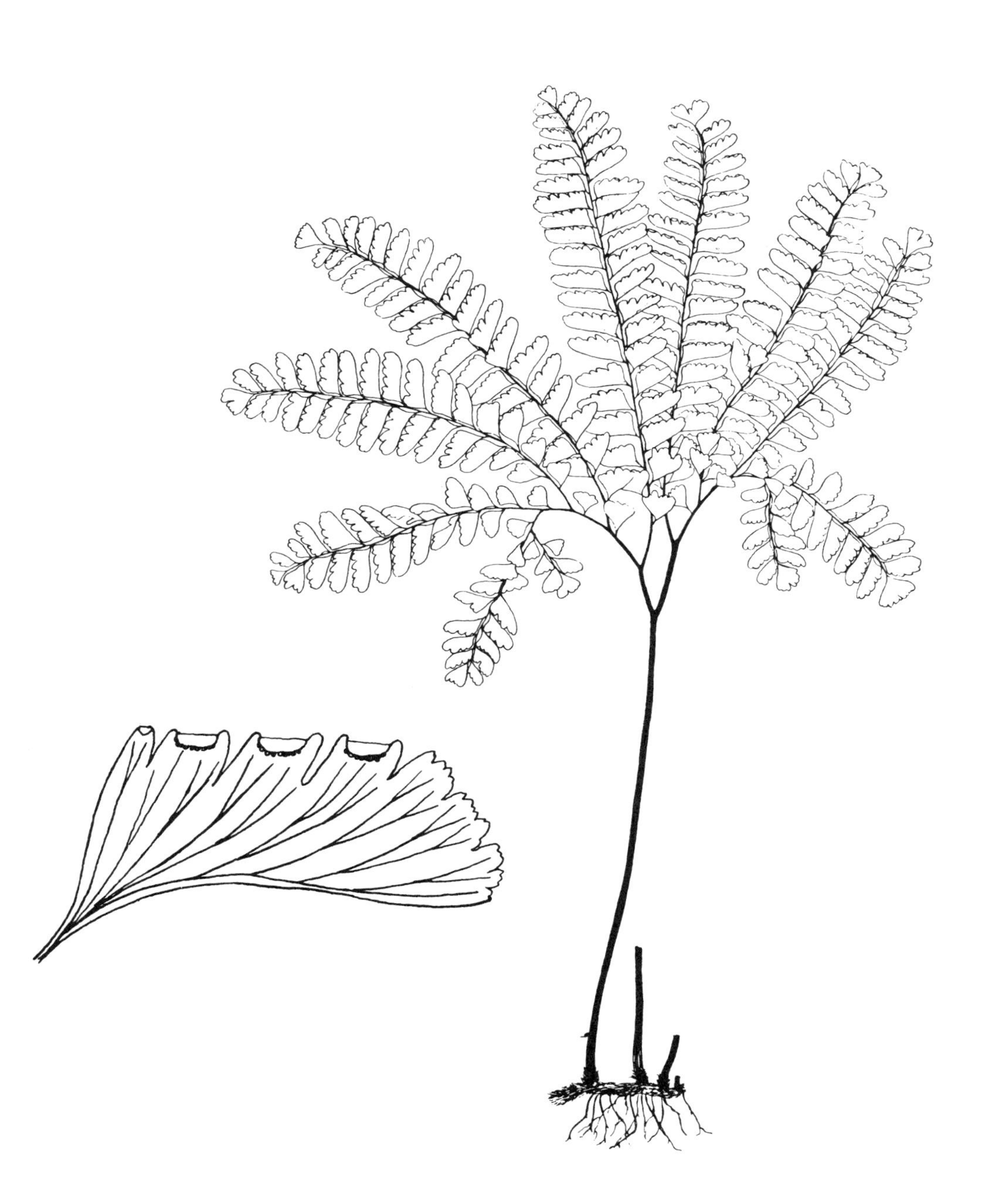

Southern Maidenhair Fern

Adiantum capillus-veneris L.

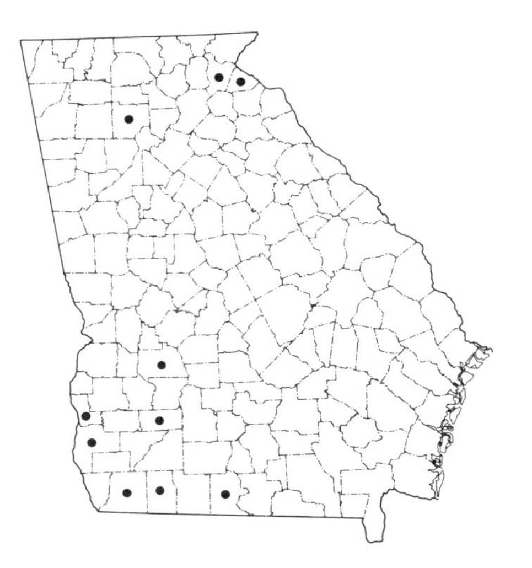

Name: Linnaeus named this fern in 1753. The epithet means "Venus's hair."

Rootstock: Slender, creeping.

Fronds: 25 to 55 cm long. Pendent or arching. Delicate. Often evergreen.

Stipe: 10 to 20 cm long. Black or dark brown. Shiny, smooth, except scaly at base.

Rachis: Dark brown. Smooth, slender, and brittle. Continuous and not forked, but noticeably zigzag.

Blade: 15 to 35 cm long and 10 to 20 cm wide. Widest near base. Triangular. Bipinnate to tripinnate.

Pinnae: Alternate, distant. Lanceolate to ovate. Compound near base, becoming simple near apex. Pinnules fan-shaped to rhomboid. 0.5 to 2 cm long. Incised. Sterile pinnae sharp-toothed.

Sori: On underside of pinnules covered by reflexed irregular lobes of margins.

Habitat: Usually restricted to wet limestone rocks, moist cliffs, limestone sinks and bluffs, and wet, rocky river banks.

Range: Tropical and warm temperate regions of the world. Southern and western sections of North America. Found in limestone sink regions of southwestern Georgia and in a few sites in the mountainous area of the Piedmont.

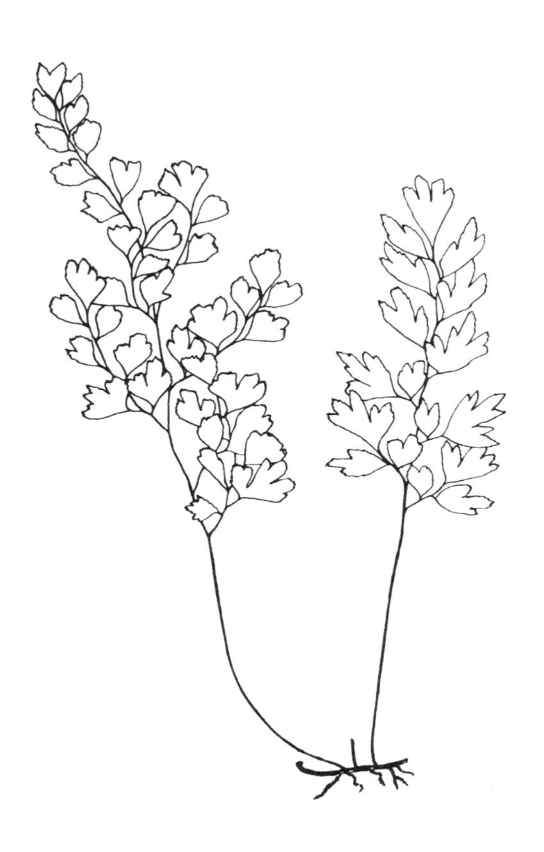

Alabama Lipfern

Cheilanthes alabamensis (Buckl.) Kunze

Name: Swartz established the genus in 1806, the name *Cheilanthes* derived from the Greek words meaning "lip" and "flower," in allusion to the marginal sori. The common name, "lipfern," refers to this character. This species was first found on limestone rocks along the Tennessee River in northern Alabama by S. B. Buckley, who placed it in *Pteris* in 1843. Kunze transferred it to *Cheilanthes* in 1847.

Rootstock: Short, slender, with narrow, orange-brown scales.

Fronds: To 50 cm long, but mostly 10 to 30 cm. Small, in loose cluster. Evergreen.

Stipe: 5 to 15 cm long. Shiny black. Naked, with a few scales at base.

Rachis: Shiny black. Upper side with short hairs.

Blade: 5 to 15 cm long and 2 to 4 cm wide. Narrow oblong. Widest at about third pair of pinnae. Apex long, acuminate. Bipinnate to bipinnate-pinnatifid. Surfaces naked or with scattered short hairs beneath.

Pinnae: Mostly alternate, distant, and almost sessile. Ascending, acute angle with rachis of 50 to 60 degrees. Ovate-lanceolate with acute apex. Pinnules nearly entire.

Sori: Marginal. Almost continuous under reflexed marginal indusia.

Habitat: Crevices and ledges of sheltered limestone rocks. Around rocks in rather dry woodlands.

Range: Frequent to common in the southern United States, northern Mexico, and Jamaica. In Georgia, in a few northern counties.

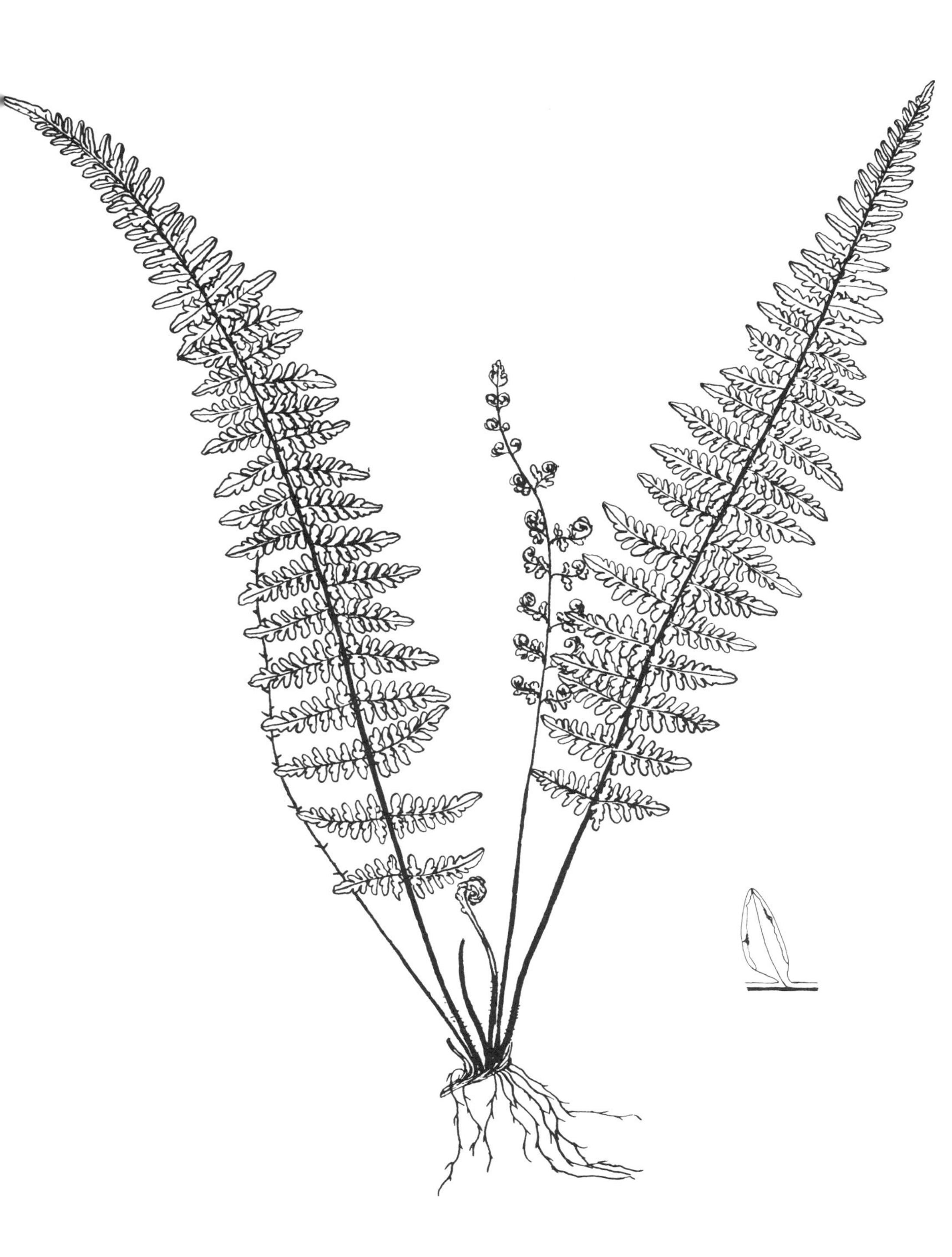

Hairy Lipfern

Cheilanthes lanosa (Michx.) D. C. Eaton

Name: Michaux discovered this fern in the southeastern United States and placed it in *Nephrodium* in 1803. Eaton transferred it to *Cheilanthes* in 1859. The specific epithet, *lanosa,* means "woolly."

Rootstock: Slender, short-creeping. Covered with brown, narrow, toothed scales.

Fronds: 20 to 40 cm long. Loosely clumped. Dark green, evergreen. Curl up when dry but revive with rain. Fertile fronds often break off in winter, while shorter, broader sterile fronds often persist.

Stipe: 8 to 15 cm tall. Dark brown. Copiously hairy. Brittle, breaks off easily.

Rachis: Dark brown. Hairy.

Blade: 12 to 25 cm long and 2 to 4 cm wide. Oblong-lanceolate, broadest near middle. Bipinnate-pinnatifid. Hirsute with lax, rusty hairs, sparse above, dense below.

Pinnae: Mostly opposite and ascending. Widely spaced below, crowded above. Lanceolate.

Sori: Continuous along margins of lobes of pinnules, with margins reflexed. Confluent at maturity, giving lower surface of pinnule a solid brown appearance.

Habitat: Non-calcareous rocky slopes, outcrops, ledges, cliffs, and rocky stream banks exposed to the sun and wind.

Range: Frequent throughout the eastern and southern United States. Common in Georgia from the granite region of the Piedmont northward.

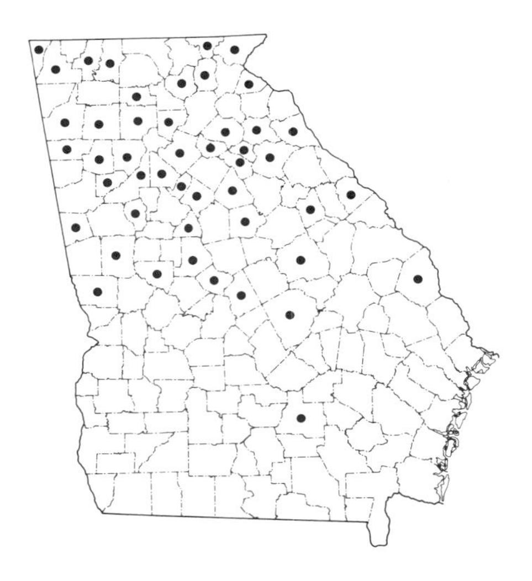

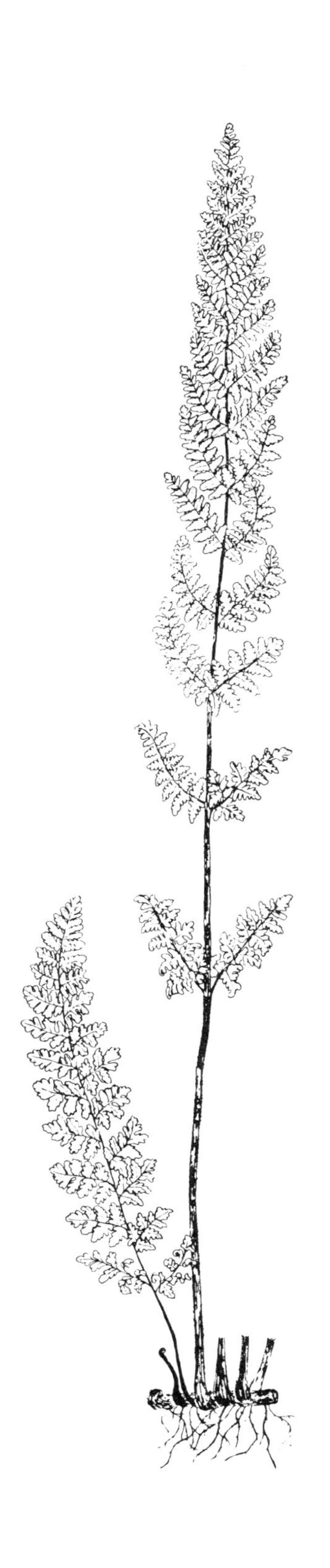

Woolly Lipfern

Cheilanthes tomentosa Link

Name: Link's 1833 name, *tomentosa,* means "densely woolly," and like the common name refers to the matted, whitish brown wool that covers the back of the leaf blade.

Rootstock: Short-creeping, stout, with brown, linear scales with a central dark stripe.

Fronds: 25 to 50 cm long. Tufted. Bright green, evergreen. Curl up when dry but become fresh and green again with rain.

Stipe: 12 to 25 cm long. Brown, densely covered with long, rusty-silvery hairs and narrow scales.

Rachis: Brown. Densely hairy and scaly.

Blade: 13 to 25 cm long and 3 to 5 cm wide. Oblong to linear. Bipinnate-pinnatifid to tripinnate. Very hairy above. Densely matted with hairs below. Hairs silvery white when young, becoming brownish with age.

Pinnae: Alternate and strongly ascending. Lower, short, ovate to triangular, 2 to 3 cm long and 1 to 2 cm wide. Upper, ovate to ovate-lanceolate, about 2 cm long and 1 cm wide. Pinnules round.

Sori: Marginal, covered by reflexed leaf margin.

Habitat: Granite and sandstone rocks, on more or less sheltered ledges.

Range: Common in the southern United States and northern Mexico. In Georgia, in numerous northern counties.

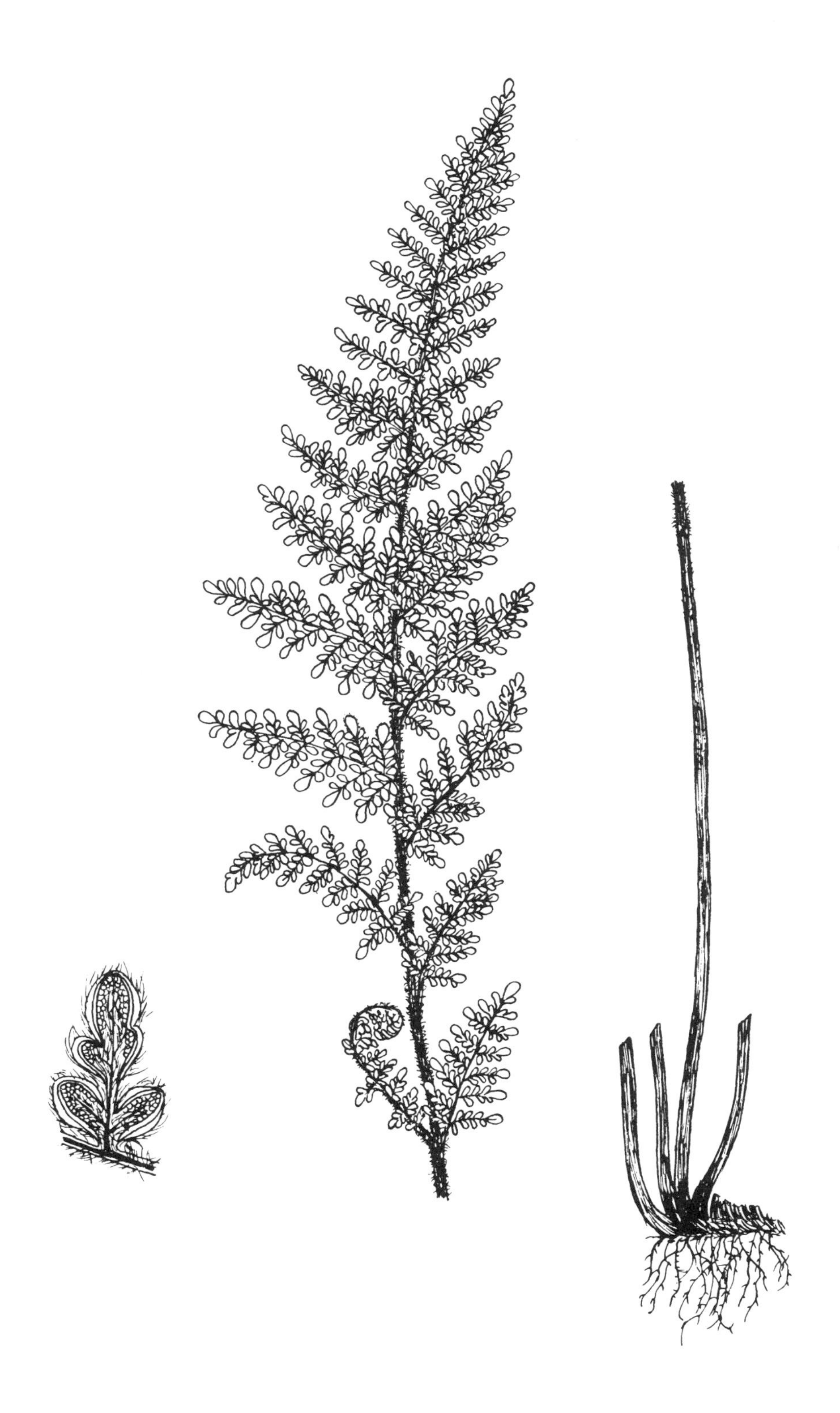

Wavy Cloak Fern

Cheilanthes sinuata (Lag. ex Sw.) Domin

Name: In 1806 the Spanish botanist Mariano Lagasca y Segura named this fern *Acrostichum sinuatum* from a specimen collected by Olof Peter Swartz, a Swedish pteridologist. The German pteridologist Georg Kaulfuss transferred it to *Notholaena* in 1824. Then in 1913 the Czechoslovakian botanist Karel Domin moved it to *Cheilanthes*. *Notholaena* means "spurious cloak" and *sinuata* "strongly waved," hence the common name.

Rootstock: Short-creeping. Stout with linear, chestnut brown, finely toothed scales.

Fronds: 15 to 45 cm long. Clustered. Long, narrow.

Stipe: 3 to 5 cm long. Round, stout. Covered with white scales.

Blade: 12 to 40 cm long and 2 to 3 cm wide. Narrow, linear. Pinnate-pinnatifid.

Pinnae: Alternate. Oblong with short petioles. Cut one-third to one-half way to midvein with 3 to 6 pairs of lobes per pinna. Upper surface with a few white dissected scales; lower surface covered with brown or white hairy scales.

Sori: Continuous along reflexed margins.

Habitat: Limestone slopes and rocks.

Range: Frequent in Central and South America and the southwestern United States. Found in one site in Georgia in 1980.

Remarks: This fern has not otherwise been found in the United States east of Texas. Its disjunct appearance in Meriwether County, Georgia, is interesting and unusual. Here is an excellent example of the long-distance dispersal of some fern spores.

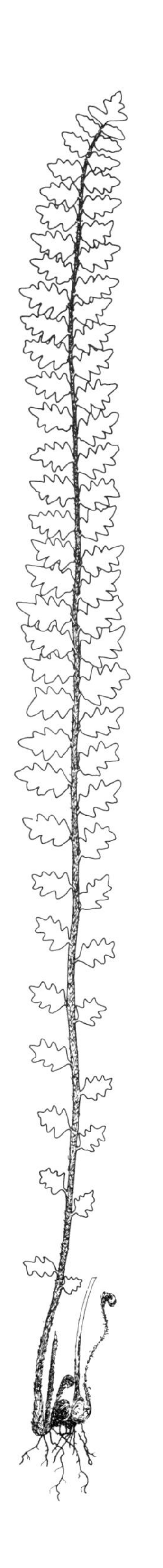
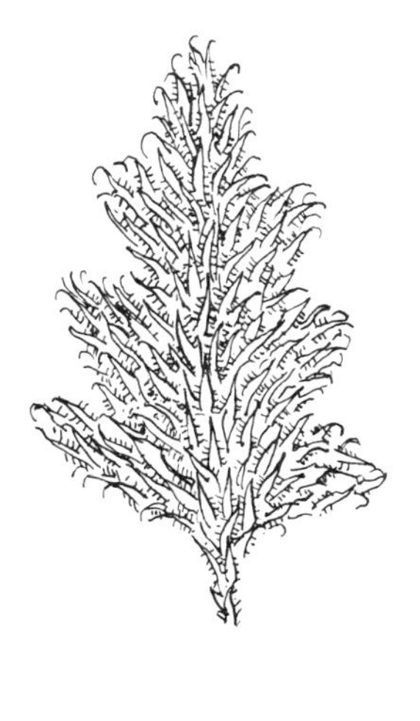

Purple Cliffbrake

Pellaea atropurpurea (L.) Link

Name: John Clayton discovered this fern on the Rappahannock River in Virginia in the early 1700s, and in 1753 Linnaeus named it *Pteris atropurpurea* ("dark-purple") in reference to the color of the leaf stalk. In 1841 Link proposed a new genus, *Pellaea*, and included this fern in it. *Pellaea* is from a Greek word meaning "dark-colored," referring to the leaf stalk. It is also commonly called "hairy cliffbrake."

Rootstock: Short-creeping. Scales rust brown, tan at tip of rhizome.

Fronds: 20 to 50 cm long. Closely bunched. Gray-green and leathery. Evergreen. Fertile fronds larger than sterile ones with narrower and more pointed divisions.

Stipe: 10 to 20 cm long. Round, dark purple to black, hairy.

Rachis: Dark purple to black. Hairy.

Blade: 10 to 30 cm long and 5 to 10 cm broad. Narrowly triangular. Bipinnate to tripinnate. Variable with divisions more complex at base.

Pinnae: Distant, almost opposite. Two to five pairs of basal pinnules and a long terminal one. Fertile pinnules mostly oblong.

Sori: Submarginal, continuous, protected by reflexed leaf margin, which opens completely when the spore cases are discharged.

Habitat: Rocks, crevices, or ledges, usually calcareous. Often in exposed situations on cliffs, as well as in more sheltered places, including brick and stone walls.

Range: Widely distributed in temperate North America. In Georgia, common in the northwestern part of state, and locally elsewhere.

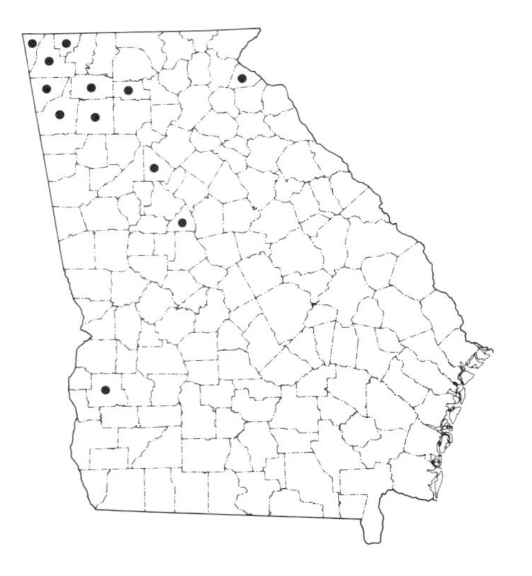

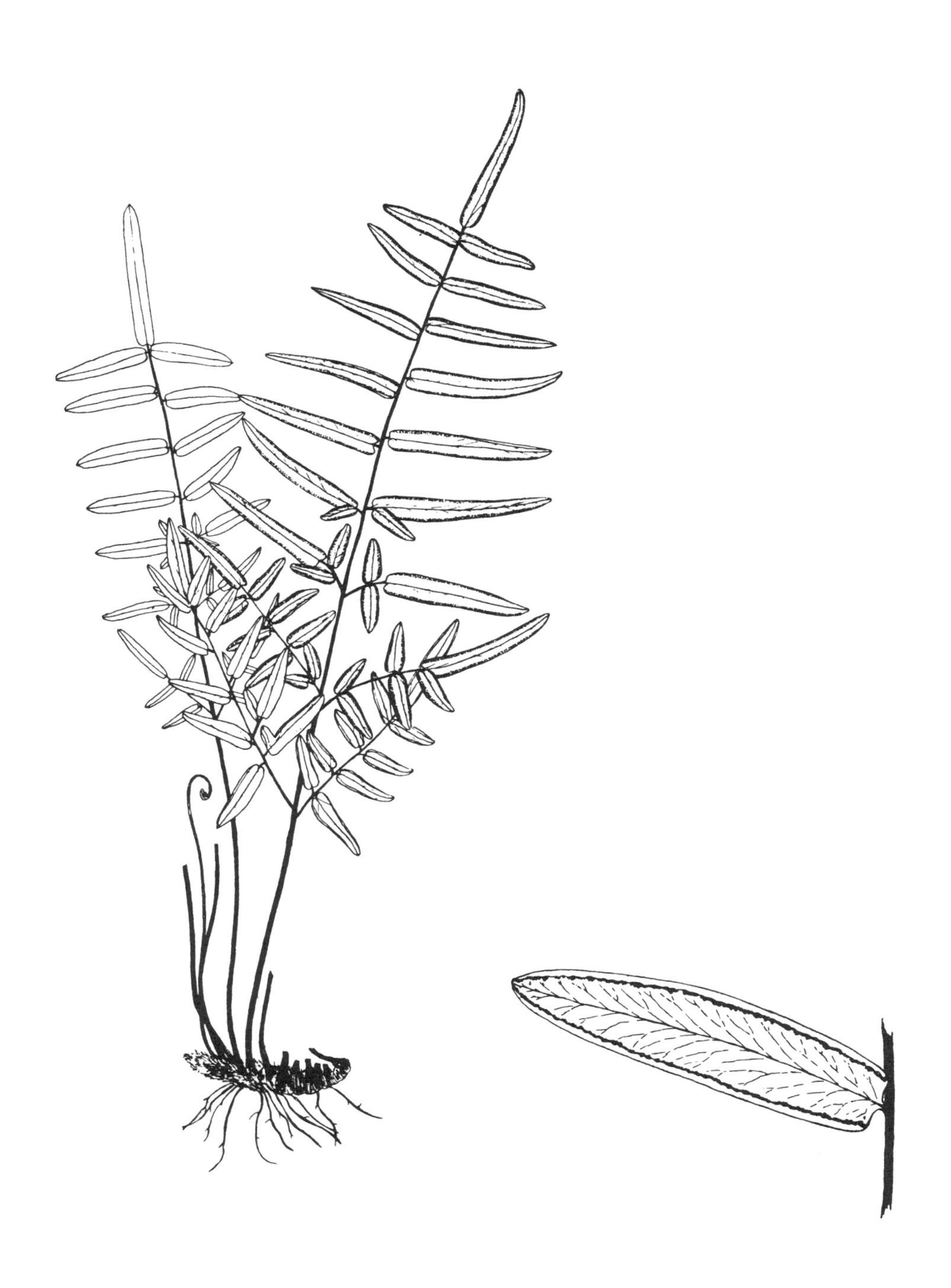

Spider Brake

Pteris multifida Poir.

Name: *Pteris,* an ancient Greek name for "fern," was used by Linnaeus as a generic designation in 1753. Poiret added this species in 1804. *Multifida* means "split into many divisions" and refers to the divided leaves. It is also called "wall fern." In Charleston, South Carolina, it is known as the "Huguenot fern" from having been found there in a Huguenot cemetery in 1868.

Rootstock: Short-creeping.

Fronds: 25 to 60 cm long. Clustered. Not evergreen, but new fronds produced throughout the year.

Stipe: 10 to 20 cm long. Smooth.

Rachis: Smooth. Winged.

Blade: 15 to 40 cm long and 12 to 25 cm broad. Oblong-triangular. Pinnate.

Pinnae: Long oblong. 3 to 7 opposite pairs plus long-oblong terminal one. Upper pinnae undivided and extend down along rachis forming a wing. Lowest divided in one or two pairs of large pinnules. Sterile pinnules broader than fertile ones and have margins finely toothed.

Sori: Submarginal, protected by reflexed margin.

Habitat: Grown as a houseplant. Waste ground, rocky woods, and masonry crevices.

Range: A native of Asia. Occurs sporadically from New York to California, and fairly well-established in the southeastern United States. Found in old walls in scattered locations in Georgia.

Ladder Brake

Pteris vittata L.

Name: Linnaeus named this fern in 1753. *Vittata* means "striped" and refers to the linear sori along the edges of the fertile pinnae.

Rootstock: Short-creeping, stout, scaly.

Fronds: 30 to 90 cm long. Tufted, erect or spreading.

Stipe: 5 to 30 cm long. Densely fine-scaled.

Rachis: Densely fine-scaled.

Blade: 25 to 60 cm long and 13 to 25 cm broad. Broadest above middle. Pinnate.

Pinnae: Oblong-lanceolate. 12 to 20 pairs strongly ascending, with very long terminal pinna. Lower pinnae relatively short, broad, and further apart. Sterile pinnae sharply toothed.

Sori: Linear, submarginal, under reflexed margins of pinnae.

Habitat: Masonry mortar and limestone, and also rocky woodlands and pinelands.

Range: Native of Asia. Escaped throughout the southeastern United States. In Georgia, only in Echols County.

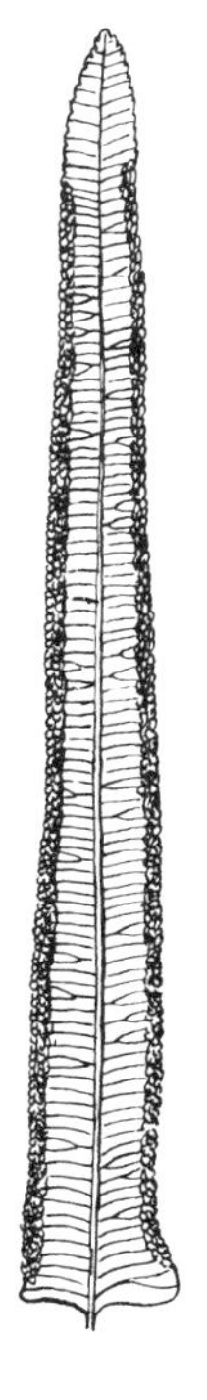

Shoestring Fern

Vittaria lineata (L.) J. E. Smith

Name: In 1753 Linnaeus named this fern *Pteris lineata* from a Santo Domingo specimen. The epithet means "marked with lines," referring to the parallel lines of the long, continuous sori. In 1793 Sir James E. Smith transferred it to his newly proposed *Vittaria,* the generic name meaning "stripe."

Rootstock: Short-creeping. Copiously scaled, scales iridescent with long, hairlike tips.

Fronds: 30 to 60 cm long and 2 to 3 cm wide. Epiphytic, clustered, pale green. Long-pendent, lustrous, leathery, resembling thick grass blades.

Sori: In two grooves running length of blade parallel to midvein. No indusia, but leaf margins inrolled to cover sori.

Habitat: Epiphytic on trunks of rough-barked trees and palmettos in moist woods. In Georgia, in rock crevices.

Range: Tropical America and common in Florida. Found in two Georgia counties, although believed to be extinct in one now.

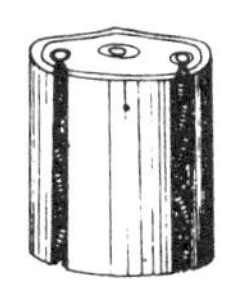

Hay-Scented Fern

Dennstaedtia punctilobula (Michx.) Moore

Name: André Michaux discovered this fern, probably in Canada, and in 1803 named it *Nephrodium punctilobulum.* In 1857 Thomas Moore transferred it to *Dennstaedtia,* a genus named by Bernhardi in 1801 in honor of the German botanist August Wilheim Dennstaedt. *Punctilobula,* meaning "having small, pointed lobes," refers to the appearance of the pinnules. "Hay-scented" comes from the alfalfa-like fragrance of the fronds.

Rootstock: Wiry, long-creeping, hairy.

Fronds: 35 to 75 cm long, but may reach 100 cm. Yellow-green, arching, deciduous. Finely cut and feathery. Produced singly at intervals. Smell like new-mown hay.

Stipe: 10 to 30 cm long. Shining, light reddish brown, darker at base. Sparsely hairy.

Rachis: Yellow-brown. Covered with fine, light-colored hairs.

Blade: 25 to 45 cm long and 13 to 25 wide. Ovate-oblong, widest near bottom. Bipinnate to bipinnate-pinnatifid. Thin-textured and easily broken. Veins below finely hairy.

Pinnae: Alternate. Lanceolate. Barely stalked. Pinnules oblong, sessile, finely cut into toothed lobes.

Sori: Marginal, in notches of pinnules. Borne in distinctive, cuplike indusia.

Habitat: Uplands in moderate shade, or moist or dry, wooded hillsides. Along edges of woods and streams and shaded roadside banks.

Range: Eastern temperate North American southward to Arkansas and northern Georgia. Common in the mountains of northern Georgia southward to Fulton and DeKalb counties.

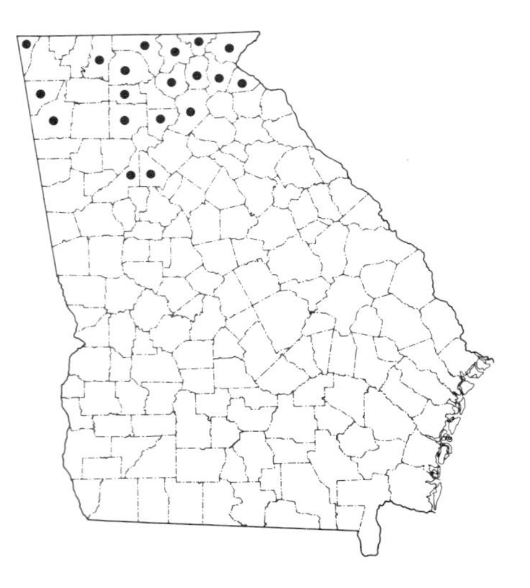

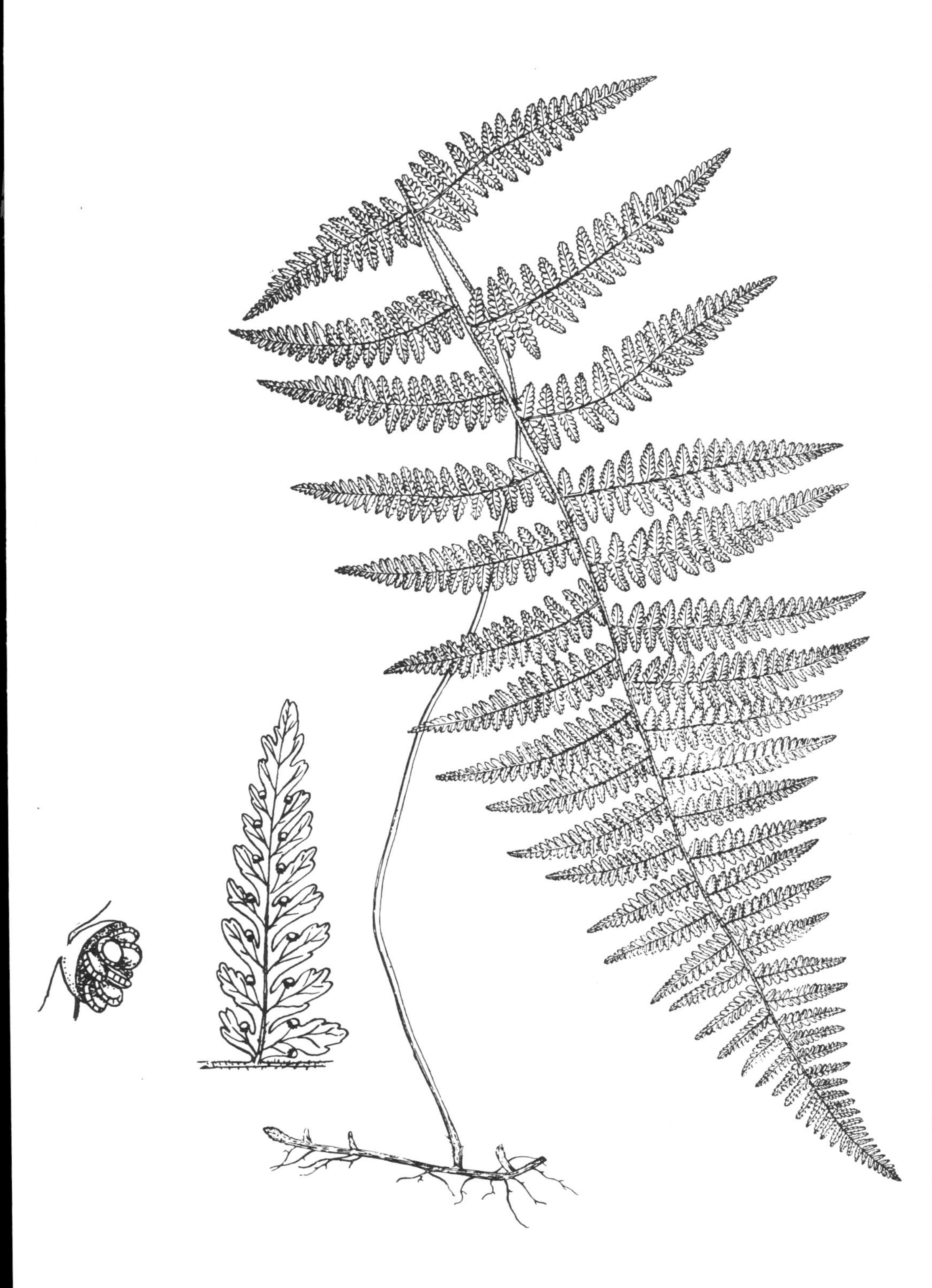

Bracken Fern

Pteridium aquilinum (L.) Kuhn

Name: Linnaeus named the bracken fern of Europe *Pteris aquilina* in 1753, and Kuhn transferred it to *Pteridium* in 1879. *Pteridium* is another name for fern, and *aquilinum* means "like an eagle," referring to the fact that the fiddleheads uncurl in three sections like the opening of an eagle's claw.

Rootstock: Long-creeping, hairy, deep in soil.

Fronds: 45 to 90 cm long. Produced singly in rows. Coarse, deciduous, rapidly spreading, ternately divided. Produced early and throughout season so that dead fronds and new green ones may be seen on same plant simultaneously.

Stipe: 20 to 40 cm long. Woody, smooth. Green, turning brown.

Rachis: Smooth, woody.

Blade: 25 to 50 cm long and 35 to 75 cm broad. Broadly triangular. Reflexed almost horizontal to ground. Bipinnate to tripinnate.

Pinnae: Oblong with distinctive long, narrow, blunt tip. Pinnules mostly long and narrow, but variable in cutting and shape.

Sori: Linear along margins of leaflets. Covered by reflexed edges of leaflets.

Habitat: Rather dry situations. Open roadsides, thinly wooded flats and hillsides, in full or partial sun.

Range: Almost worldwide. Found in practically every county of Georgia.

Remarks: Two varieties of *Pteridium aquilinum* grow in Georgia. Var. *latiusculum* (Desv.) Underw. (meaning "broadish") is sparsely hairy beneath and has pinnule tips about 0.6 cm wide and less than 4 times as long as wide. It is found mostly in northern Georgia. Var. *pseudocaudatum* (Clute) Heller (meaning "tailed-like") is naked beneath and has pinnule tips mostly about 0.3 cm wide and more than six times as long as wide. Var. *pseudocaudatum* is most common in the Coastal Plain. Both varieties, however, as well as various intermediate plants, are found throughout the state.

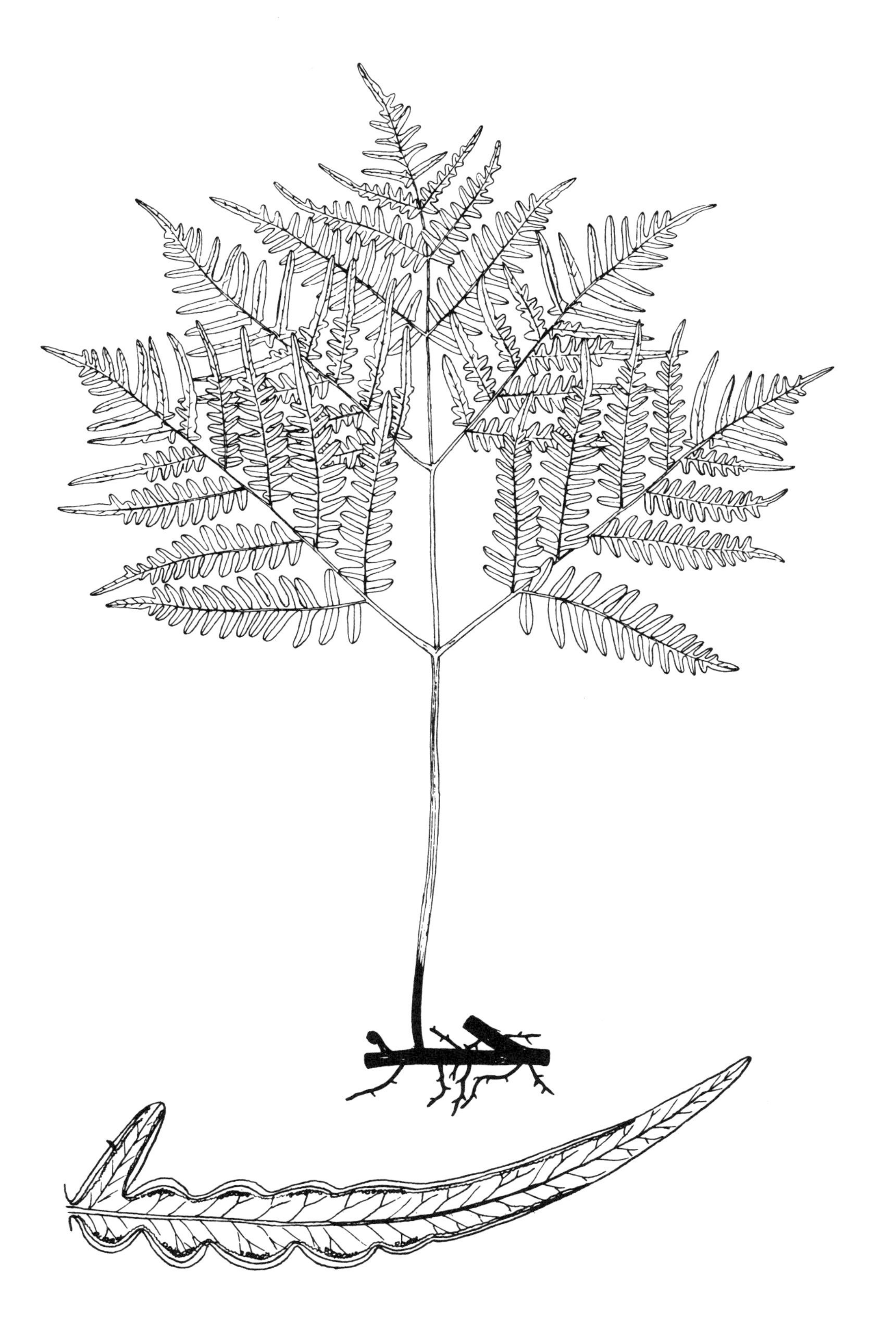

Appalachian Filmy Fern

Trichomanes boschianum Sturm

Name: In 1753 Linnaeus proposed the genus *Trichomanes,* meaning "a tangled mass of hair," from an ancient Greek name for a now unknown small, hairy fern. In 1861 Sturm named this taxon in honor of R. B. van den Bosch, a nineteenth-century botanist who wrote about the filmy fern. The name "filmy fern" is applied to members of the genus because of the delicate texture of the fronds. "Bristle fern" is also used, referring to the bristlelike projection to which sporangia are attached.

Rootstock: Long-creeping, wiry, dark-hairy.

Fronds: 8 to 15 cm long. Distant, produced singly in rows. Lax, arching or pendent. Thin, delicate, filmy and translucent. Only one cell thick. Light green, evergreen.

Stipe: 2 to 3 cm long. Green, winged.

Rachis: Green, winged.

Blade: 6 to 12 cm long and 2 to 4 cm wide. Pinnate-pinnatifid.

Pinnae: Much dissected. Irregularly cut or lobed. Blunt tipped.

Sori: Sporangia borne on bristlelike extensions of fertile leaflets, emerging from tubular cuplike indusia.

Habitat: Moist non-calcareous rocks, crevices, and cave mouths, in situations kept perpetually humid and deeply shaded.

Range: Rare, found only in the eastern United States. In Georgia, only in Rabun County, where it is known from three different sites.

Peters' Filmy Fern

Trichomanes petersii A. Gray

Name: T. M. Peters found this fern in northwestern Alabama and Asa Gray named it in his honor in 1853. It is also called "dwarf filmy fern" and "bristle fern."

Rootstock: Long-creeping, black, hairy, delicate.

Fronds: Tiny, only 1 or 2 cm long. Produced singly. Masses of fronds often forming a green covering over rocks where the plant grows. Dark green, evergreen. Resemble leaves of broad-leaved mosses. Thin, translucent, only one cell thick.

Stipe: None, or very short.

Blade: 1 to 2 cm long and less than 0.5 cm wide. Undivided. Veins branched, fanlike. Long, wiry, wide-spaced hairs along edges.

Sori: Sporangia on short, stiff bristle projecting from indusia. Indusia funnel-shaped with flared opening on leaf terminal.

Habitat: Moist, sheltered rocks, mostly sandstone, where air is perpetually moist. Rarely on shaded bases of tree trunks.

Range: Rare, confined to the southeastern United States. In Georgia, in several widely scattered localities.

Blackstem Spleenwort

Asplenium resiliens Kunze

Name: In 1753 Linnaeus established the genus *Asplenium* using an ancient Greek word for a plant supposed to cure the diseases of the spleen. In 1842 Martens and Galeotti named this fern *Asplenium parvulum* from a Mexican plant. As this name had already been used, Kunze, in 1844, gave the plant its present designation, which means "springy." The common name, referring to a distinguishing characteristic, is appropriate.

Rootstock: Short, erect, covered with a mass of brown roots, persistent stipe bases and long, black scales.

Fronds: 6 to 30 cm long. Small, tufted, evergreen. Leathery, dark green. Intermixed with old fronds and stipes.

Stipe: 1 to 5 cm long. Black, shiny.

Rachis: Black, shiny.

Blade: 5 to 25 cm long and 1 to 2.5 cm wide. Lanceolate to narrowly oblong. Widest above middle. Pinnate.

Pinnae: 0.5 to 1.5 cm long and 0.2 to 0.5 cm wide. Mostly opposite and well spaced with extremely short stalks. Oblong with blunt tips. Lower pinnae somewhat triangular. Superior auricles. Margins smooth or nearly so.

Sori: Short, oblong. Medial or submarginal. Often on all pinnae of fertile frond. Indusia thin, laterally attached.

Habitat: Small crevices of shaded calcareous rocks.

Range: Various parts of subtropical America northward to Arizona, Missouri, and southern Pennsylvania. Common in the limestone valleys of northwestern Georgia, and also found in the Coastal Plain.

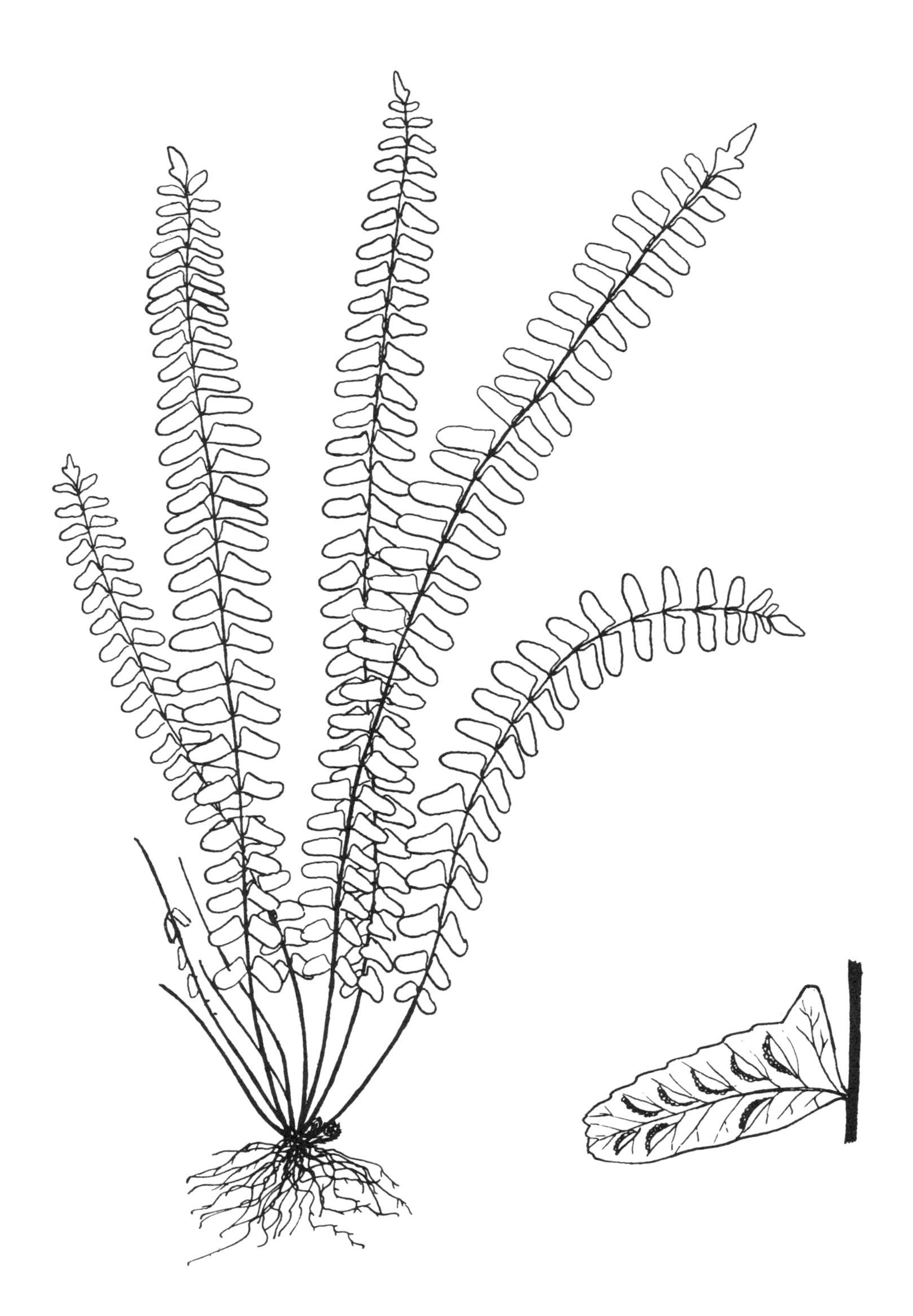

Varicolored Spleenwort

Asplenium heterochroum Kunze

Name: This taxon was discovered in Florida a number of years before it was distinguished from *A. resiliens* and named by Kunze in 1834. The epithet means "varying in color" and refers to the rachis.

Rootstock: Stout with inconspicuous short, dark brown, narrow scales.

Fronds: 12 to 37 cm long. Erect, tufted, evergreen.

Stipe: 2 to 7 cm long. Dark brown or blackish. Narrowly winged.

Rachis: Grades upward from black to brown with green wings.

Blade: 10 to 30 cm long and 2 to 3 cm wide. Linear. Broadest in middle or just above; tapers to both apex and base. Pinnate.

Pinnae: Mostly opposite. Variable in shape. Larger pinnae elliptical with rounded apex, wedgelike base, auricled on upper side. Lower pinnae gradually reduced, ovate-rhomboid to fan-shaped. Margins finely incised. Veins simple or once-forked.

Sori: Linear-elliptical. Medial. Indusia laterally attached.

Habitat: Rocky hammocks, calcareous rocks, and limestone sinks.

Range: Frequent in Florida, Bermuda, and the West Indies. In Georgia, only in Camden and Colquitt counties, currently the northernmost known sites in the United States.

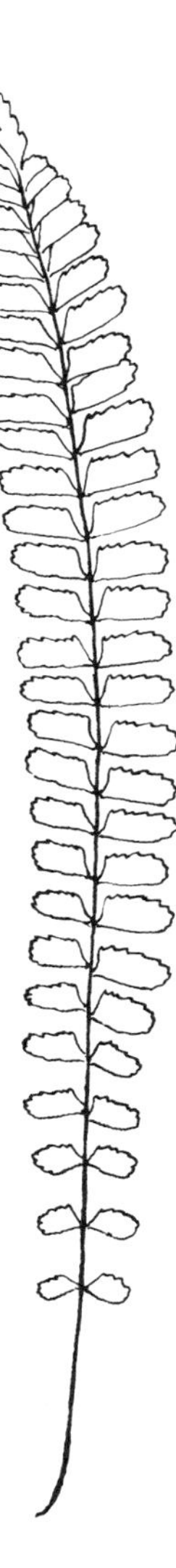

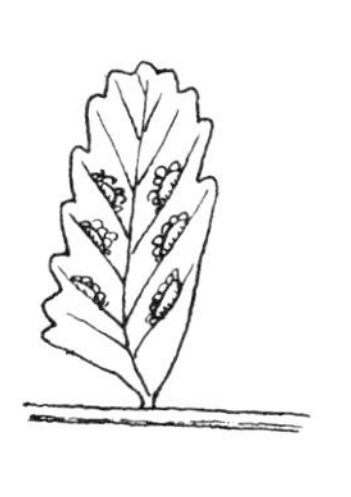

Wagner's Spleenwort

Asplenium heteroresiliens Wagner

Name: From a specimen collected in Florida in 1960, W. H. Wagner called this fern *Asplenium heterochroum* × *resiliens* in 1963, and in 1966 he described it as a new species, *A. heteroresiliens.*

Rootstock: Stout, erect, with dark scales.

Fronds: 10 to 30 cm long. Erect, tufted, evergreen.

Stipe: 2 to 5 cm long. Dark brown to black.

Rachis: Dark brown to black.

Blade: 8 to 25 cm long and 2 to 3 cm wide. Broadest above middle. Firm and leathery.

Pinnae: Triangular-ovate to oblong-ovate. Superior auricles. Lower third of pinnae somewhat descending with small auricles on lower side. Margins variable but usually shallowly cut with coarse teeth.

Sori: Linear-elliptical. Submarginal. Indusia laterally attached.

Habitat: Rocky hammocks, calcareous rock crevices, and limestone sinks.

Range: Only in the Carolinas, Georgia, and Florida. In Georgia, only in Lee County in the Upper Coastal Plain.

Remarks: *Asplenium heteroresiliens* is a fertile hybrid of *A. resiliens* and *A. heterochroum.* This is the only pteridophyte on Georgia's list of endangered plants.

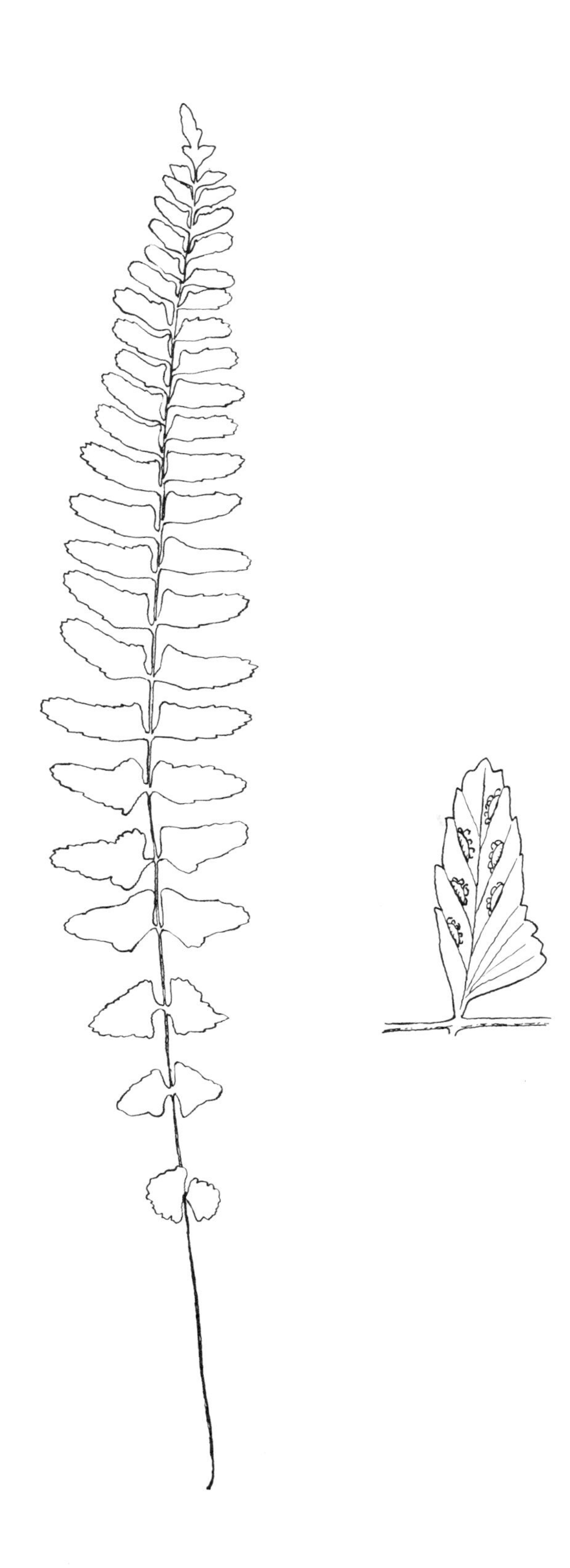

Maidenhair Spleenwort

Asplenium trichomanes L.

Name: Linnaeus described this from European material in 1753. The epithet, meaning "a tangled mass of hair," alludes to the plant's delicate texture, like that of the filmy ferns of the genus *Trichomanes.* The dark stipe and rachis, delicate texture, and shape of the pinnae suggest the maidenhair ferns of the genus *Adiantum,* hence the common name.

Rootstock: Slender and much branched. Rootstock, roots, scales, and old stipe bases form a small, dense, dark mass that may reach about 3 cm long.

Fronds: 5 to 20 cm long. Tufted, erect to sinuous, evergreen. Thin and delicate in texture.

Stipe: 2 to 5 cm long. Dark purple-brown.

Rachis: Dark purple-brown.

Blade: 3 to 15 cm long and about 1 cm wide. Pinnate. Tapers slightly to both ends.

Pinnae: Tiny (only about 0.5 cm in length and width). Lower mostly opposite, upper slightly alternate. Rounded or semioblong. Barely stalked and not auricled. Margins with round teeth.

Sori: Few, medial, elongate. Confluent with age. Indusia laterally attached.

Habitat: Pockets and crevices, usually in deep shade, in both calcareous and non-calcareous rocks. Tolerates most rock types.

Range: One of the relatively few plants occurring in identical form in North America, Europe, and Asia. Widely distributed in North America except in the central plains states. Fairly common throughout the mountain areas of northern Georgia.

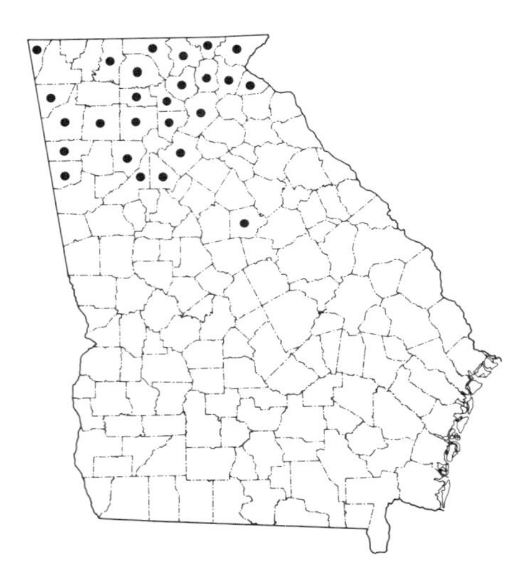

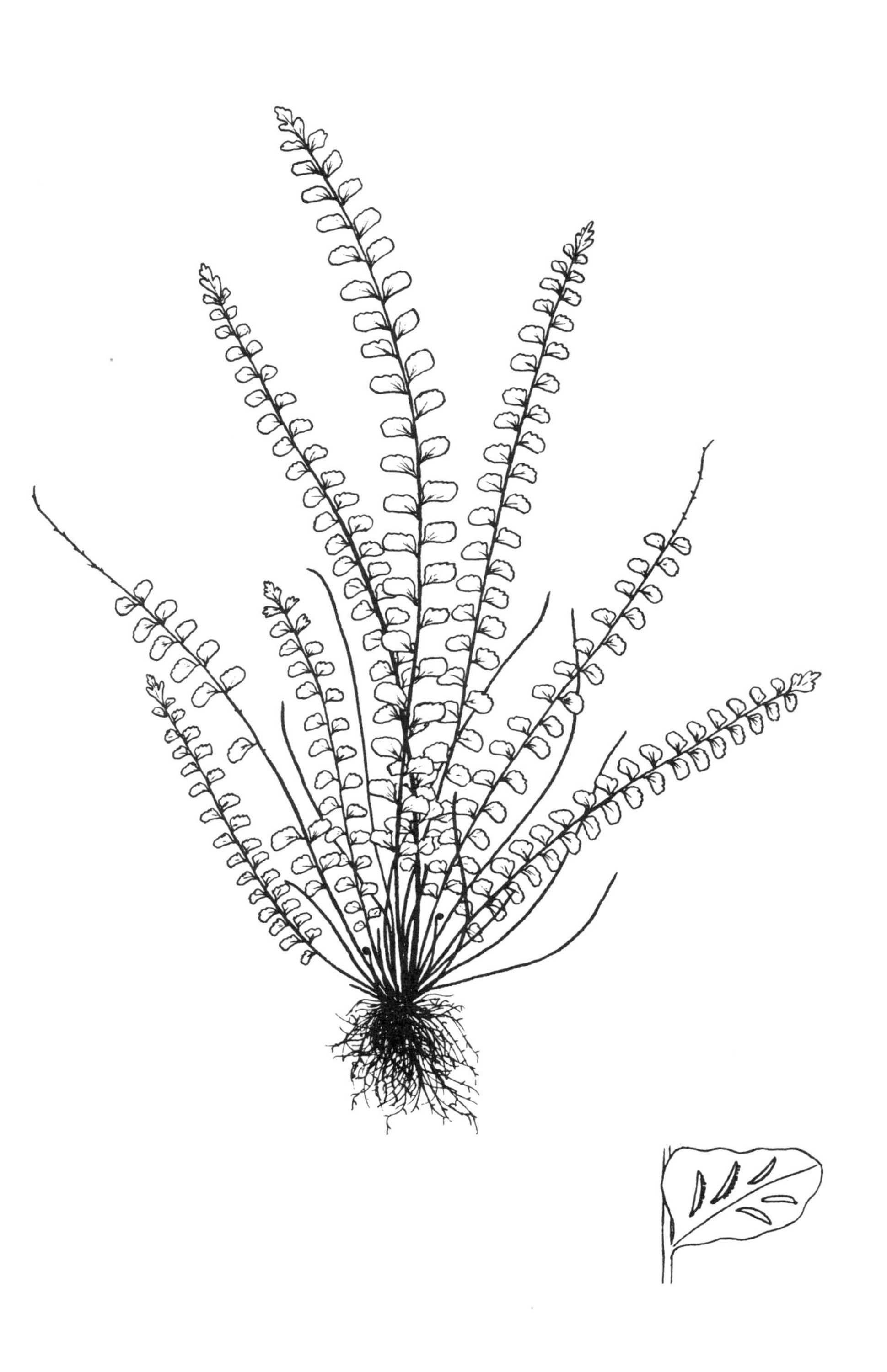

Ebony Spleenwort

Asplenium platyneuron (L.) Oakes ex D. C. Eaton

Name: In 1753 Linnaeus named this *Acrostichum platyneuron* from a Virginia specimen. Oakes transferred it to *Asplenium*, the name being formally published by D. C. Eaton in 1878. The epithet, meaning "broad-veined," is inappropriate and due to a mixup of taxa. It is commonly called "ebony spleenwort" from the epithet *A. ebeneum* given by William Aiton in 1789 because of its nearly black stipe and rachis. It is also called "brownstem spleenwort."

Rootstock: Thick, short, creeping.

Fronds: Fertile, 15 to 40 cm long. Centrally placed in tufts. Erect, tardily deciduous. Sterile, and partially fertile, 8 to 12 cm long, spreading, lying almost flat, evergreen.

Stipe: On fertile frond, 3 to 6 cm long. On sterile frond, 1 to 3 cm long. Shining dark brown, often nearly black.

Rachis: Shiny dark brown.

Blade: Fertile, 12 to 34 cm long and 2.5 to 5 cm wide. Sterile, 7 to 9 cm long and 1 to 2 cm wide. Oblong, tapering to both ends. Pinnate.

Pinnae: To 2.5 cm long and 0.5 cm wide. Alternate, oblong, with superior auricles. Sessile, the pinna bases overlapping the rachis. Edges incised variously. Lower pinnae on fertile frond sterile.

Sori: One row on each side of pinna midvein. Elongate, becoming confluent and dark brown. Indusia laterally attached.

Habitat: Wooded slopes, open woods, rocky places. Invades masonry. Best in well-drained, subacidic soil.

Range: Abundant throughout much of eastern North America. Extremely common throughout Georgia except for the southeastern Coastal Plain.

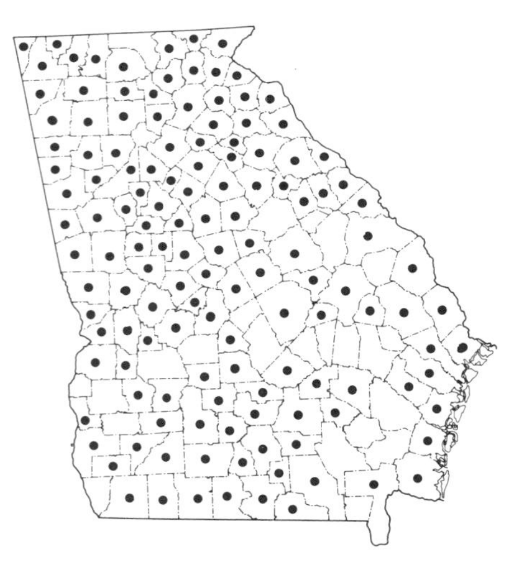

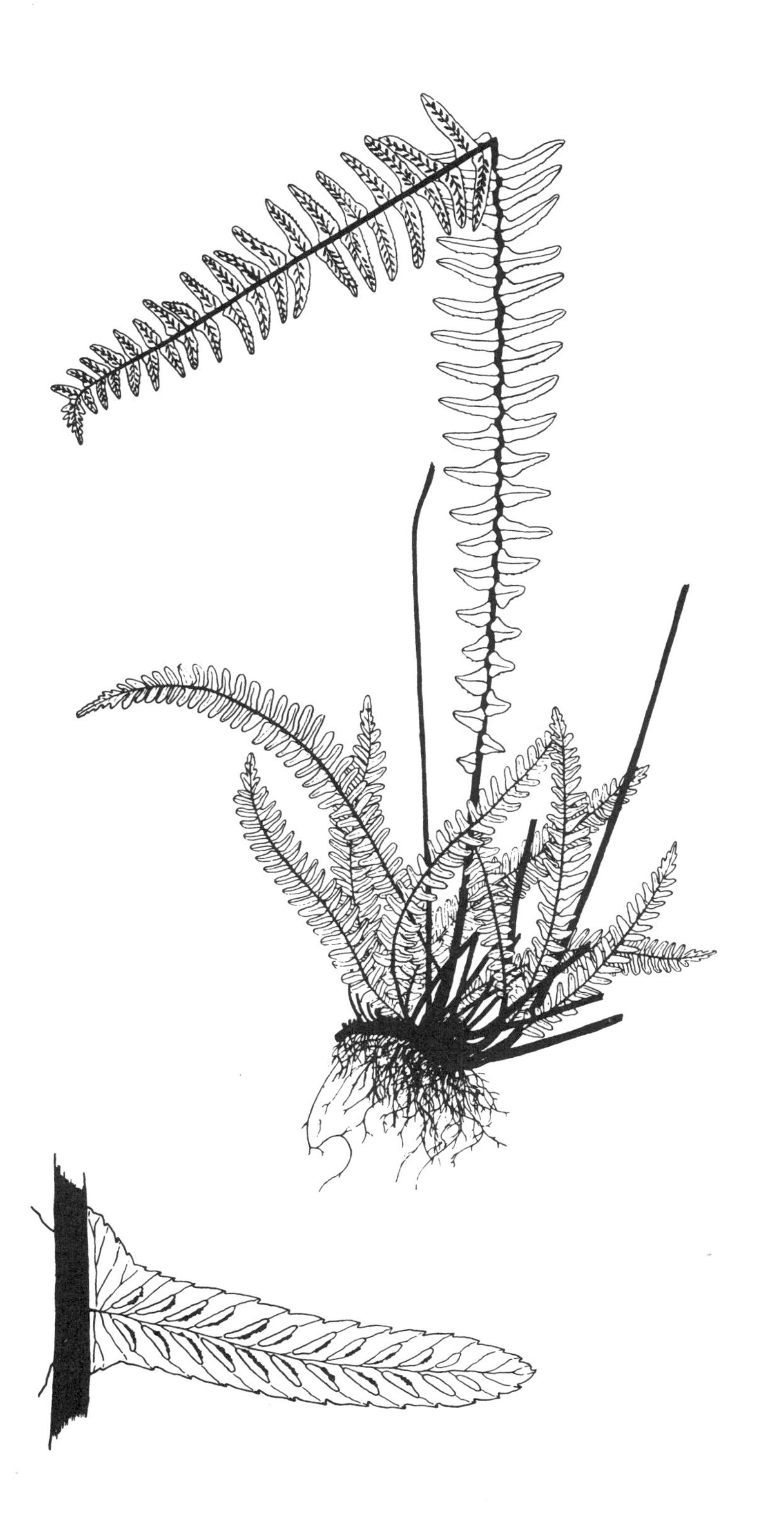

Bradley's Spleenwort

Asplenium bradleyi D. C. Eaton

Name: Eaton named *A. bradleyi* in 1873 in honor of F. H. Bradley, who discovered it in eastern Tennessee. It is also commonly called "cliff spleenwort" because of its habitat.

Rootstock: Small, dark, wiry, creeping, with persistent old stipe bases.

Fronds: 8 to 20 cm long. Few, tufted, evergreen.

Stipe: 2 to 6 cm long. Shining dark brown.

Rachis: Lower one-half to two-thirds shining dark brown. Upper part green.

Blade: 6 to 14 cm long and 1 to 4 cm wide. Oblong-lanceolate to oblong. Pinnate with pinnatifid apex.

Pinnae: Alternate. Basal triangular becoming oblong-lanceolate to oblong with lobed apex. Basal 1 or 2 pairs of pinnae shorter than those above. Superior auricles. Margins serrate to jagged.

Sori: Short, numerous, medial. Linear indusia laterally attached.

Habitat: Crevices of overhanging cliffs of non-calcareous rocks.

Range: Eastern New York to Oklahoma and southeastward to central Georgia. Very local and not usually abundant. Scattered counties in northern and central Georgia.

Remarks: A fertile hybrid of *A. montanum* and *A. platyneuron*. Hybridizes with *A. pinnatifidum* to produce the sterile *Asplenium* × *gravesii*, and with *A. montanum* to produce *A.* × *wherryi*.

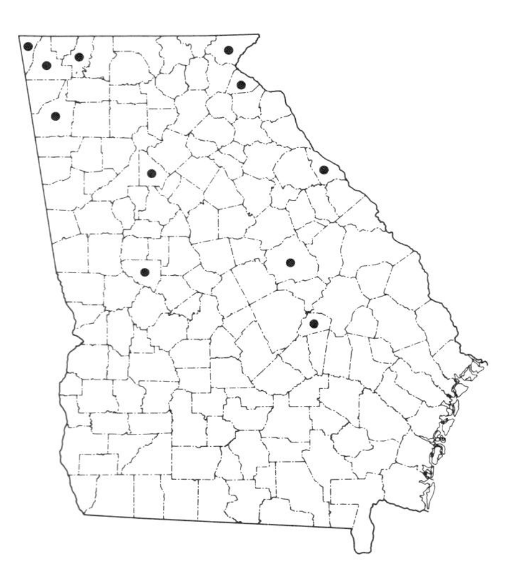

Lobed Spleenwort

Asplenium pinnatifidum Nutt.
Asplenosorus pinnatifidus (Nutt.) Mickel

Name: Thomas Nuttall named this species in 1818, the first plants having been found along the Schuylkill River in Pennsylvania. Both the common name and the Latin epithet refer to the lobed fronds.

Rootstock: Short and erect or horizontal in rock crevices.

Fronds: 5 to 18 cm long. Few, clustered, evergreen.

Stipe: 2 to 6 cm long. Brown below, green above.

Rachis: Green and smooth.

Blade: 3 to 12 cm long and 1 to 2 cm wide. Simple. Long, triangular with long, slender tip. Pinnately lobed in rounded segments, variously asymmetrical. Veins mostly free.

Sori: Scattered and variable in form. Often confluent. Indusia distinct, laterally attached.

Habitat: Dry, shaded crevices of noncalcareous rocks and boulders. Soil subacidic.

Range: One of the "Appalachian spleenworts," a group confined to the eastern United States. Ranges from Pennsylvania to Illinois, southward to Oklahoma and middle Georgia. Found in about 10 northern and central Georgia counties.

Remarks: A fertile hybrid of *Asplenium montanum* and *Camptosorus rhizophyllus*. *A. pinnatifidum* itself hybridizes with other *Aspleniums* to form sterile hybrids—with *A. bradleyi* to form *A.* × *gravesii*; with *A. platyneuron* to form *A.* × *kentuckiense*; and with *A. montanum* to form *A.* × *trudellii*.

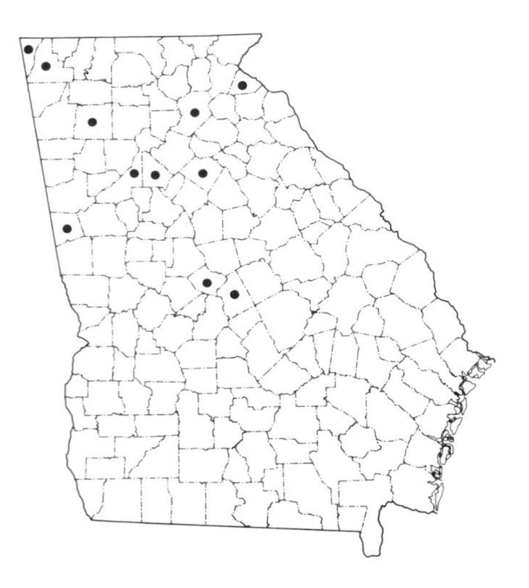

Trudell's Spleenwort

Asplenium × trudellii Wherry
Asplenosorus × trudellii (Wherry) Mickel

Name: Edgar T. Wherry named this plant in 1925 in honor of Harry W. Trudell, an amateur botanist of Philadelphia who collected a specimen in Lancaster County, Pennsylvania, in 1920.

Rootstock: Short, slender, creeping. Covered with old, persistent stipe bases.

Fronds: 7 to 17 cm long. Clustered, few, erect to arching, evergreen.

Stipe: 2 to 7 cm long. Dark brown lower half, green upper.

Rachis: Green. Slender at base, winged above.

Blade: 5 to 10 cm long and about 2 cm wide. Lanceolate to narrowly deltoid with long, acuminate apex. Pinnate at base with 2 to 5 pairs of pinnae widely spaced on definite, slender stalks. Upper part pinnatifid, often deeply lobed. Widest at base.

Pinnae: About 0.8 cm long and wide. Lower deltoid-ovate, upper ovate-lanceolate. Margins coarsely toothed.

Sori: Short, scattered, medial. Indusia laterally attached.

Habitat: Acidic soil in sheltered crevices of sandstone and other non-calcareous rocks.

Range: Sporadic from northern Alabama and Georgia to New Jersey, Pennsylvania, and Ohio. First found in Georgia in 1917 by E. W. Graves near Trenton, Dade County. Only other known site in Georgia is in Fulton County.

Remarks: *Asplenium × trudellii* is a sterile hybrid of *A. pinnatifidum* and *A. montanum.*

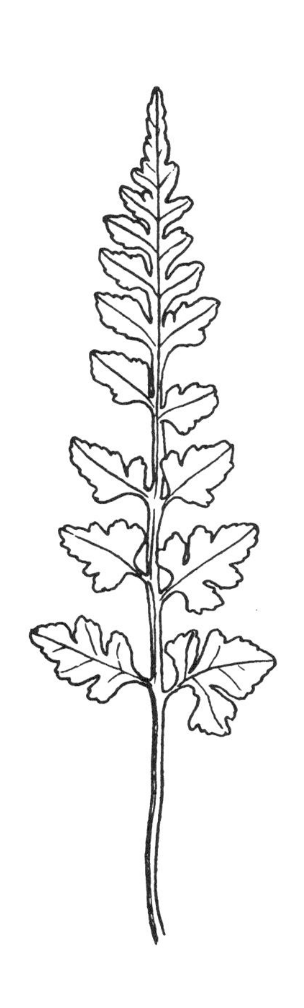

Graves' Spleenwort

Asplenium × *gravesii* Maxon
Asplenosorus × *gravesii* (Maxon) Mickel

Name: E. W. Graves discovered this fern on Sand Mountain near Trenton, Dade County, Georgia, in 1917, and the following year Maxon named it in his honor.

Rootstock: Slender, creeping.

Fronds: 5 to 20 cm long. Clustered, wide-spreading, evergreen.

Stipe: 2 to 5 cm long. Chestnut brown.

Rachis: Chestnut brown at base, green above. Slightly winged.

Blade: 3 to 15 cm long and 2 to 3 cm wide. Long and narrow with long, tapering tip. Base pinnate, apex pinnatifid.

Pinnae: Ovate to triangular, broadest at base, blunt tipped. Basal 3 to 5 pairs with distinct short stalks. Margins slightly serrate.

Sori: Elongate, medial. Indusia laterally attached.

Habitat: Non-calcareous boulders, rock outcrops, cliffs, or ledges in partial shade.

Range: A rare fern with distribution limited to a few states in the eastern United States. In Georgia, found only in Dade County.

Remarks: *Asplenium* × *gravesii* is a sterile hybrid of *A. pinnatifidum* and *A. bradleyi.*

Kentucky Spleenwort

Asplenium × *kentuckiense* McCoy
Asplenosorus × *kentuckiense* (McCoy) Mickel

Name: Thomas N. McCoy first found this fern in Kentucky in 1934 and in 1936 named it after that state.

Rootstock: Short. creeping.

Fronds: 8 to 20 cm long. Tufted, erect, spreading.

Stipe: 3 to 6 cm long. Shining brown to black.

Rachis: Lower one-third to one-half brown to black, upper green and flat.

Blade: 5 to 14 cm long and 1.5 to 3 cm wide. Lanceolate with long tapering blunt tip. Pinnate for one-half to two-thirds length with pinnatifid apex.

Pinnae: 0.8 to 1.6 cm long and 0.3 to 1 cm wide. Basal pinnae deltoid progressing to oblong with pinnatifid lobes at apex. Basal 4 to 6 pairs short stalked. Lowest pair with superior auricles. Margins mostly entire.

Sori: Plentiful, covering backs of pinnae. Medial, elongate. Indusia laterally attached.

Habitat: Shaded crevices of hard rocks including sandstone and granite gneiss.

Range: Very rare. Found sparingly in the eastern United States. In Georgia, only in Stephens County and on Brown's Mountain near Macon, in Bibb County.

Remarks: *Asplenium* × *kentuckiense* is a sterile hybrid of *A. platyneuron* and *A. pinnatifidum.*

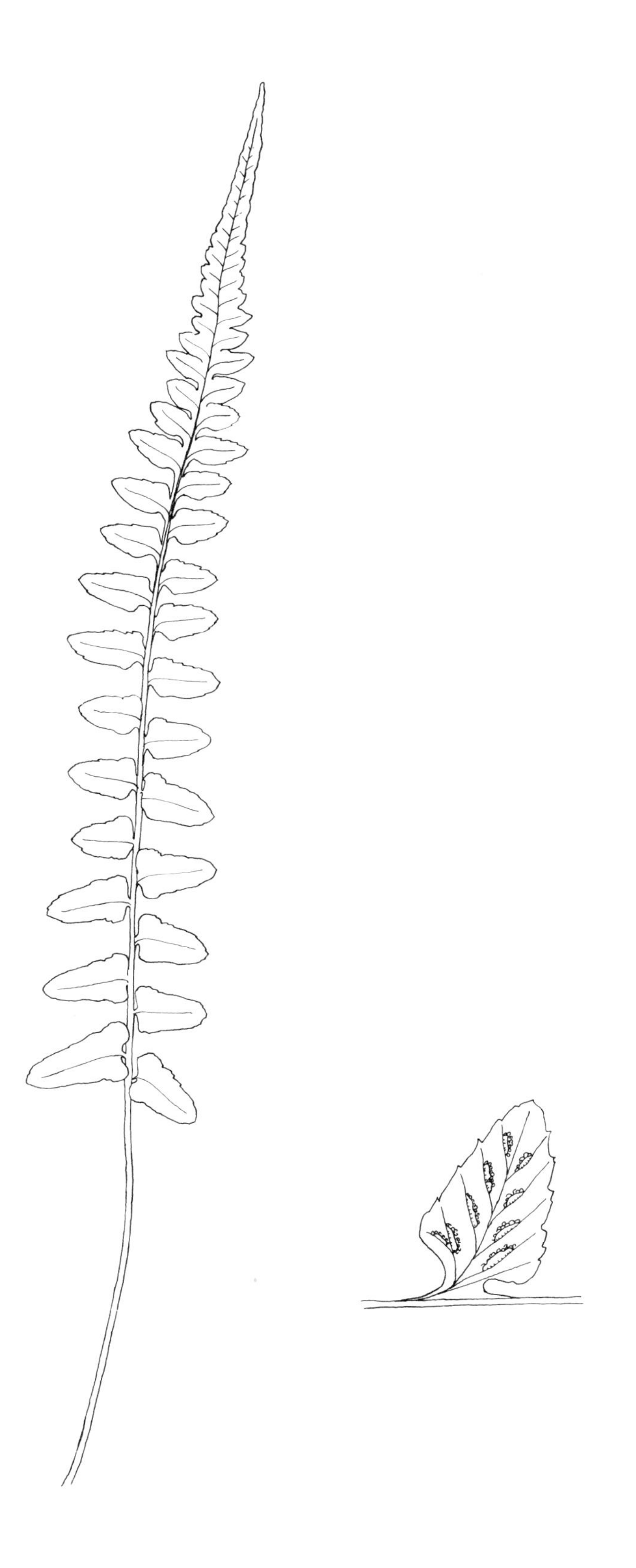

Scott's Spleenwort

Asplenium × ebenoides Scott
Asplenosorus × ebenoides (Scott) Mickel

Name: The American naturalist Robert Scott discovered and named this fern in 1865. The epithet means "resembling the taxon *ebeneum*" (ebony), another name given to *Asplenium platyneuron.*

Rootstock: Short, vertical, with black, wiry roots.

Fronds: 10 to 30 cm long. Tufted. Evergreen.

Stipe: 3 to 10 cm long. Shining purplish brown.

Rachis: Lower part brown, upper green.

Blade: 7 to 20 cm long and 2 to 5 cm wide. Long-oblong with long, tapering tip. Cut into a wide variety of patterns, irregular in outline. Pinnate at base, pinnatifid toward apex.

Pinnae: The pinnae and segments are very asymmetrical. Some are short, some long, some almost linear, some short and rounded. Some of the lower segments resemble pinna but are attached to the rachis broadly and connected by narrow green wings. Margins of pinnae and segments shallowly toothed or wavy.

Sori: Linear. Scattered at base of larger pinnae and segments. Indusia attached along one side to veins.

Habitat: Shaded, moist, calcareous rocks.

Range: Rare. Scattered locations throughout the eastern United States.

Remarks: *Asplenium × ebenoides* is a hybrid of *A. platyneuron* and *Camptosorus rhizophyllus.* In most cases it is sterile, although some rare populations are fertile.

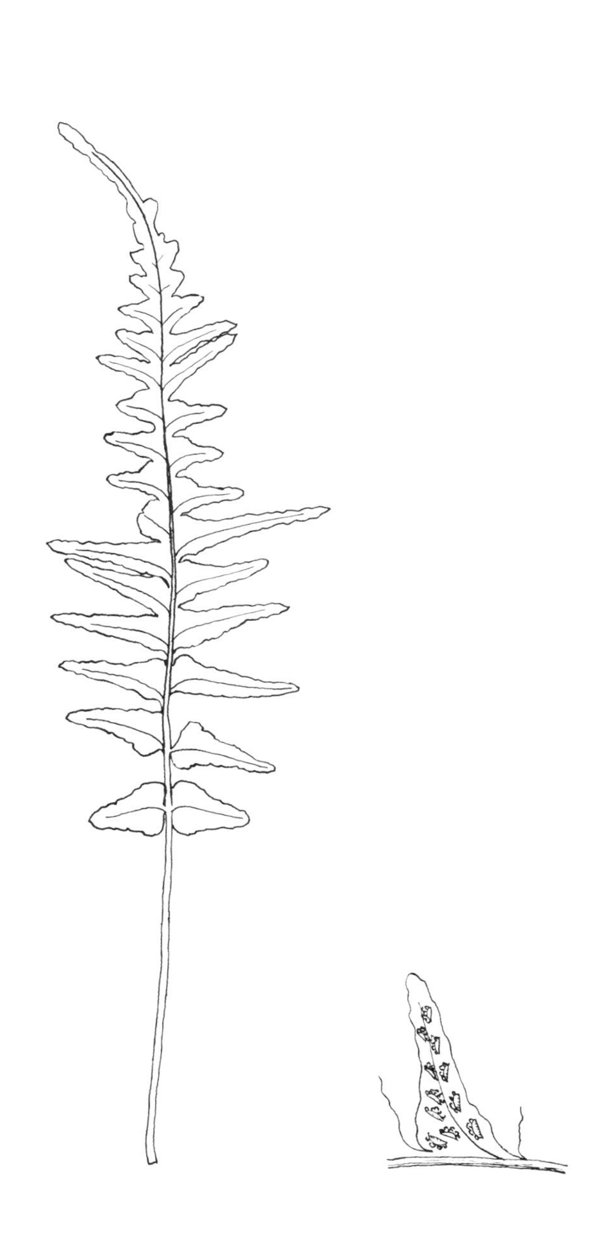

Mountain Spleenwort

Asplenium montanum Willd.

Name: Appropriately named (*montanum* means "of mountains") in 1810 by the German taxonomist Karl Willdenow for a plant from the mountains of "Carolina."

Rootstock: Short-creeping, wiry, dark. Often obscured by old stipe bases.

Fronds: 4 to 15 cm long. Numerous, usually in drooping tufts. Moderately variable in form. Delicate, bluish green, evergreen.

Stipe: 2 to 5 cm long. Dark brown below (up to one-half length, usually much less), green above.

Rachis: Broad, flat, green. Winged toward apex.

Blade: 2 to 10 cm long and 1 to 6 cm wide. Oblong-triangular. Base bipinnate, upper portion pinnatifid, lobed apex. Shiny, smooth.

Pinnae: 4 to 7 pairs. Subopposite, oblong-triangular, with short, broad stalks. Pinnate at base and pinnatifid at apex. Basal pinnae largest, 1 to 3 cm long and 0.5 to 2 cm wide. Margins vary from merely serrate to cut in irregular segments.

Sori: Few, elongate, medial, often confluent. Indusia thin, laterally attached.

Habitat: Acidic soil in shaded crevices of non-calcareous rocks.

Range: Massachusetts to Ohio and southward to Alabama and Georgia. Fairly frequent in the mountains of northern Georgia.

Wherry's Spleenwort

Asplenium × wherryi Smith, Bryant, and Tate

Name: This hybrid was first discovered by Edgar T. Wherry in New Jersey in 1935. After discovering additional plants in Kentucky, D. M. Smith, T. R. Bryant, and D. E. Tate named it in honor of Wherry in 1961.

Rootstock: Short-creeping, dark, wiry.

Fronds: 5 to 18 cm long. Tufted. Evergreen.

Stipe: 2 to 6 cm long. Shining dark brown.

Rachis: Flat, green, grooved.

Blade: 3 to 12 cm long and 1 to 6 cm wide. Lanceolate. Base bipinnate, grading to bipinnatifid and pinnatifid at apex.

Pinnae: 5 to 8 pairs. Alternate. Oblong-triangular with short, broad stalks. Pinnate at base and pinnatifid at apex. Basal pinnae largest. Pinnae 1 to 3 cm long and 0.5 to 2 cm wide. Margins serrate to cut in irregular segments.

Sori: Elongate, medial, often confluent. Indusia thin, laterally attached.

Habitat: Shaded crevices of non-calcareous rocks.

Range: A few eastern states. Fairly rare.

Remarks: A sterile hybrid of *A. montanum* and *A. bradleyi.*

Wall Rue Spleenwort

Asplenium ruta-muraria L.

Name: The epithet Linnaeus gave this fern in 1753 means "rue of the walls," rue being an aromatic Eurasian plant.

Rootstock: Vertical, slender, short. Covered with old stipe bases.

Fronds: 2.5 to 10 cm long. Tufted. Dull green. Evergreen. Dainty, inconspicuous, and unfernlike.

Stipe: 1 to 5 cm long. Green, but brown at extreme base. Glabrous except for a few short hairs.

Rachis: Green and flat.

Blade: 1.5 to 5 cm long and 1 to 4 cm wide. Oval-triangular. Bipinnate.

Pinnae: Alternate, long-stalked, composed usually of 2 to 5 wedged-shaped pinnules narrowing at base. Margins toothed or indented. The numerous, close-set veins branch 2 to 4 times and do not extend to margins of pinnules.

Sori: Medial. Vary from numerous to few. Indusia linear, attached to vein by single margin.

Habitat: Moderately shaded limestone and calcareous cliffs.

Range: Europe and Asia. Eastern North America from Ontario southward to North Carolina and Georgia and westward to Michigan and Missouri.

Walking Fern

Camptosorus rhizophyllus (L.) Link
Asplenium rhizophyllum L.

Name: Linnaeus named this fern in 1753 from an American plant. In 1833 Link established the genus *Camptosorus* and included this fern in it. *Camptosorus*, from the Greek words meaning "bent," and "sorus," refers to the irregular soral arrangement. The common name "walking fern" and the specific epithet *rhizophyllus*, meaning "root leaf," come from the ability of the plant to root at the tip of its leaves.

Rootstock: Short, slender, erect.

Fronds: 13 to 25 cm long. Clustered, evergreen. Lying on ground or arching. Undivided. Narrowly linear-deltoid with long, fine tip. Tips touch ground and root, forming new plants.

Stipe: Length variable, 1 to 9 cm, usually around 3 or 4 cm. Dark brown at base, above green and flat.

Rachis: Green and flat.

Blade: 12 to 16 cm long and 1.5 to 4 cm wide at base. Base usually heart-shaped or eared. Margins usually entire, but sometimes wavy. Veins netted.

Sori: Linear or linear-elliptical. Scattered. Indusia attached laterally along a vein.

Habitat: Sheltered rocks or walls, especially limestone.

Range: Mostly in the eastern United States. From southern Quebec to Minnesota, southward to Oklahoma and Georgia. In Georgia, primarily in the limestone regions of the northern counties.

Remarks: *Camptosorus rhizophyllus* hybridizes with several species of *Asplenium*, and these hybrids, fertile and sterile, John T. Mickel has placed in the hybrid genus *Asplenosorus*.

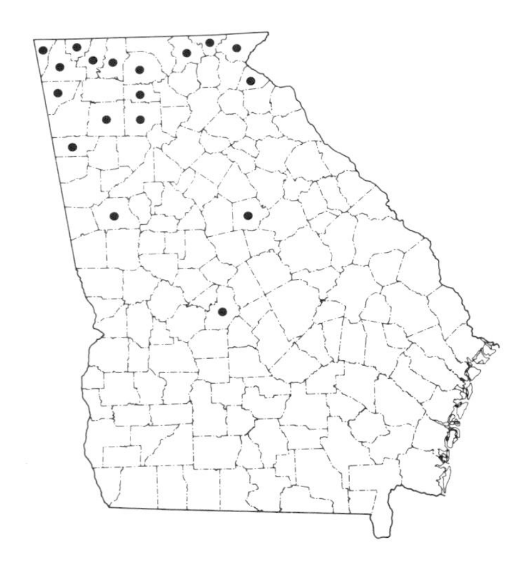

Hammock Fern

Blechnum occidentale L.

Name: In 1753 Linnaeus selected *Blechnum,* the ancient Greek name for some now unknown plant, as the name for this genus of ferns. That same year, through a slip, he named this fern *Blechnum orientale,* but corrected it in 1763 to *occidentale,* which means "western."

Rootstock: Scaly, ascending.

Fronds: 20 to 50 cm long. Clustered. Color varies from pinkish when young to reddish bronze to shiny bright green when mature.

Stipe: 10 to 25 cm long. Bronze in color. Coarse. Scaly below, hairy above.

Rachis: Covered with soft hairs.

Blade: 10 to 25 cm long and 5 to 10 cm wide. Oblong-triangular. Broadest at base, gradually tapering to apex with no distinct terminal segment. Pinnate below, pinnatifid at tip.

Pinnae: Subopposite, broadly attached at base. Linear-shaped, tapering to a somewhat blunt tip. Margins smooth or finely toothed.

Sori: In two parallel rows on each side of pinna midvein, approximately linear length of pinnae. At maturity, when indusia open, sporangia appear as two reddish brown, velvety stripes.

Habitat: Hammocks, areas of deep, rich soil and hardwood vegetation. Also, damp, open woods and loamy soil over limestone.

Range: Common in the American tropics—the West Indies, Mexico, Central and South America. Rare in the United States in Florida, Louisiana, and southern Texas. In Georgia, only in Echols County.

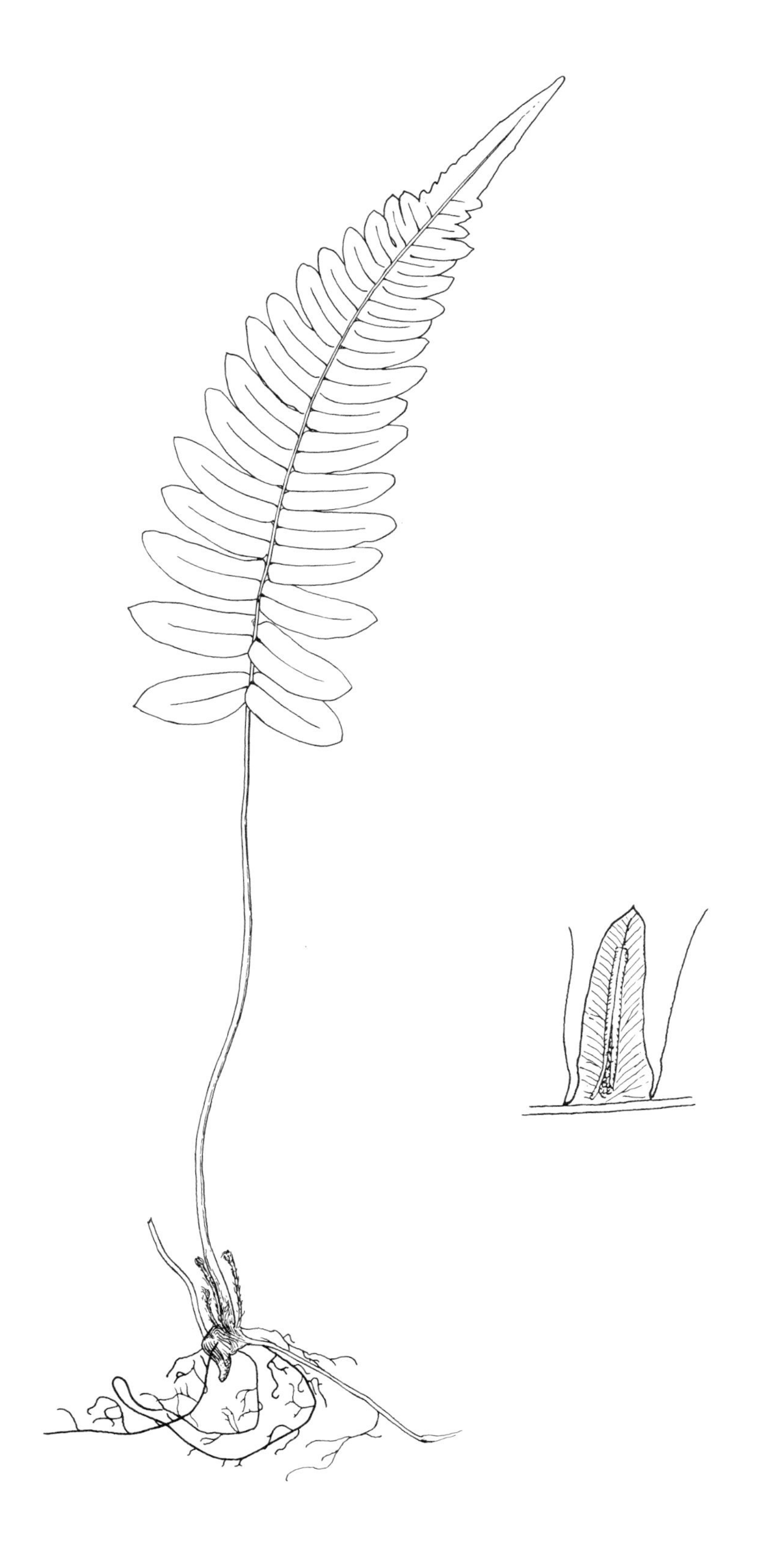

Virginia Chain Fern

Woodwardia virginica (L.) Sm.

Name: In 1771 Linnaeus named this taxon *Blechnum virginicum,* and in 1793 Sir James E. Smith transferred it to his newly founded genus *Woodwardia,* named for a friend, T. J. Woodward. The epithet means "of Virginia." This fern is also called "giant chain fern," "chain" in both common names referring to the chainlike rows of sori formed along the midveins.

Rootstock: Stout, long-creeping.

Fronds: 50 to 130 cm long. Produced in rows. Coarse, thick and firm in texture. Deciduous. Fertile and sterile alike. Fertile appearing in summer.

Stipe: 20 to 60 cm long. Shining dark brown, purplish brown below.

Rachis: Smooth. Light brown.

Blade: 30 to 70 cm long and 15 to 30 cm broad. Pinnate-pinnatifid. Ovate-deltoid, widest near base.

Pinnae: Alternate. Lanceolate with pointed tip and narrowed base. Lobes short, blunt, smooth-margined. Small veins form single row of areoles on each side of midveins, but all others are free and extend to margins of lobes.

Sori: Chainlike. Linear, in single row on each side of and parallel to midribs of pinnae and also paralleling midveins of pinnules. When mature become confluent forming a continuous line.

Habitat: Wet, swampy woods, open, swampy pinelands, and acidic bogs. Common in roadside ditches in southern Georgia, growing in full sun as long as its "feet" are wet.

Range: Bermuda. Eastern North America from Texas to Florida and northward along the coast to Nova Scotia and inland to Michigan. In Georgia, grows widely over the southern half of the state and also in scattered localities in the northern sections.

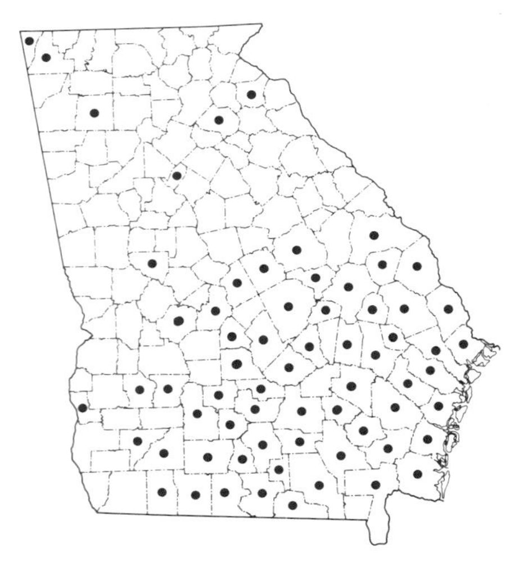

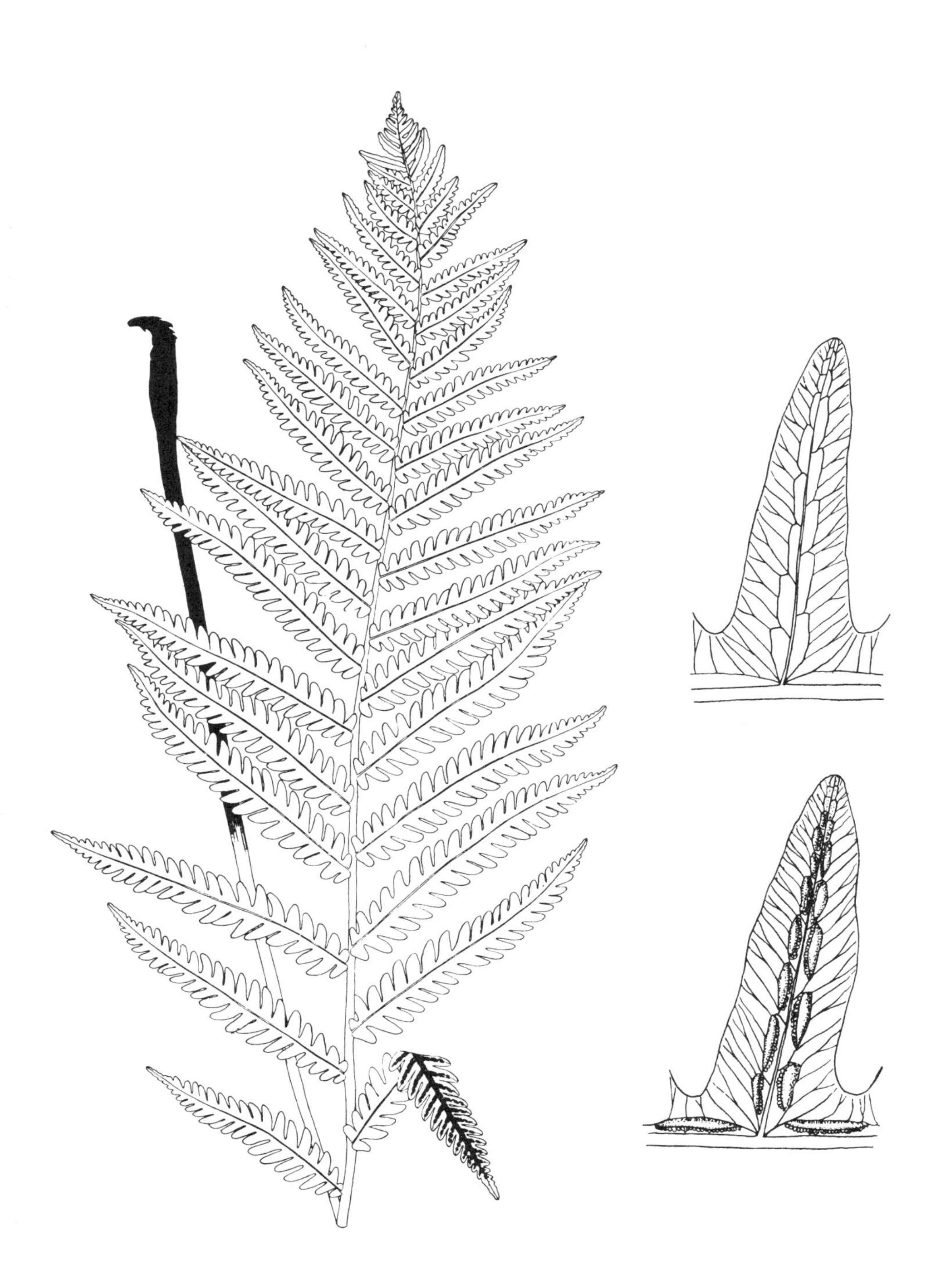

Netted Chain Fern

Lorinseria areolata (L.) Presl

Name: Linnaeus named *Acrostichum areolatum* in 1753. The epithet means "divided into small areas" and refers to the network of small veins. In 1857 Thomas Moore placed this in J. E. Smith's *Woodwardia*, and it is so included by many botanists. In 1851, however, Presl established a new genus, *Lorinseria*, including this species. The name is given in honor of the Austrian botanist Lorinser.

Rootstock: Slender, long-creeping, and branching.

Fronds: Dimorphic. Scattered. Sterile, 30 to 60 cm long. Produced in early spring, coarse, deciduous. Fertile, 50 to 80 cm long. Produced August to October, persist through winter.

Stipe: Sterile, 15 to 25 cm long. Straw-colored, dark at base. Fertile, 30 to 45 cm long. Shiny dark brown.

Rachis: Sterile, straw-colored or green. Fertile, shiny dark brown.

Sterile Blade: 15 to 35 cm long and 10 to 20 cm wide. Ovate-triangular, widest below middle. Pinnatifid. Divided into 8 to 12 pairs of alternate segments that are long-oblong, about 1.5 cm wide, with pointed tips. Lowest with a short stalk, others broadly winged at axis. Margins finely serrated. Prominent veins conspicuously netted.

Fertile Blade: 20 to 35 cm long. Segments up to 10 cm long and less than 0.5 cm wide. Alternate, widely spaced, ascending, and lax. Lower with short stalks. Upper connected basally by very narrow wing.

Sori: Long-oblong, in chainlike rows on each side of midvein and covering almost entire back side of fertile segments. Sporangia dark brown. Indusia dark brown, glabrous, elongate.

Habitat: Wet, often acidic soil of swampy woods, along streams, in rich woods and swampy depressions, and on wet, shaded rocks.

Range: Common throughout southeastern North America, especially in coastal areas. In Georgia, common in the Coastal Plain and also frequent in the northern part of the state. One of the most widespread ferns in Georgia.

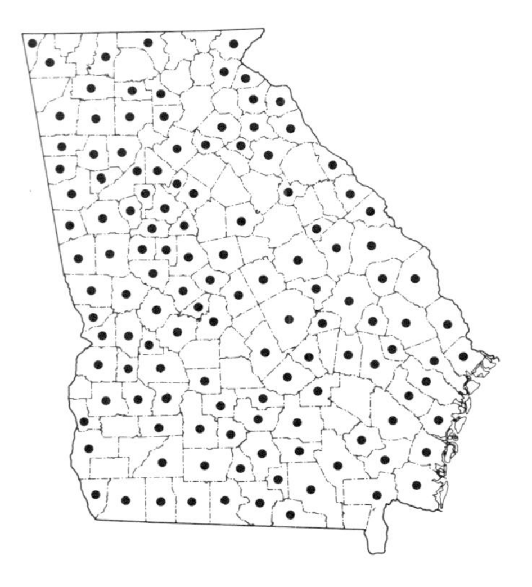

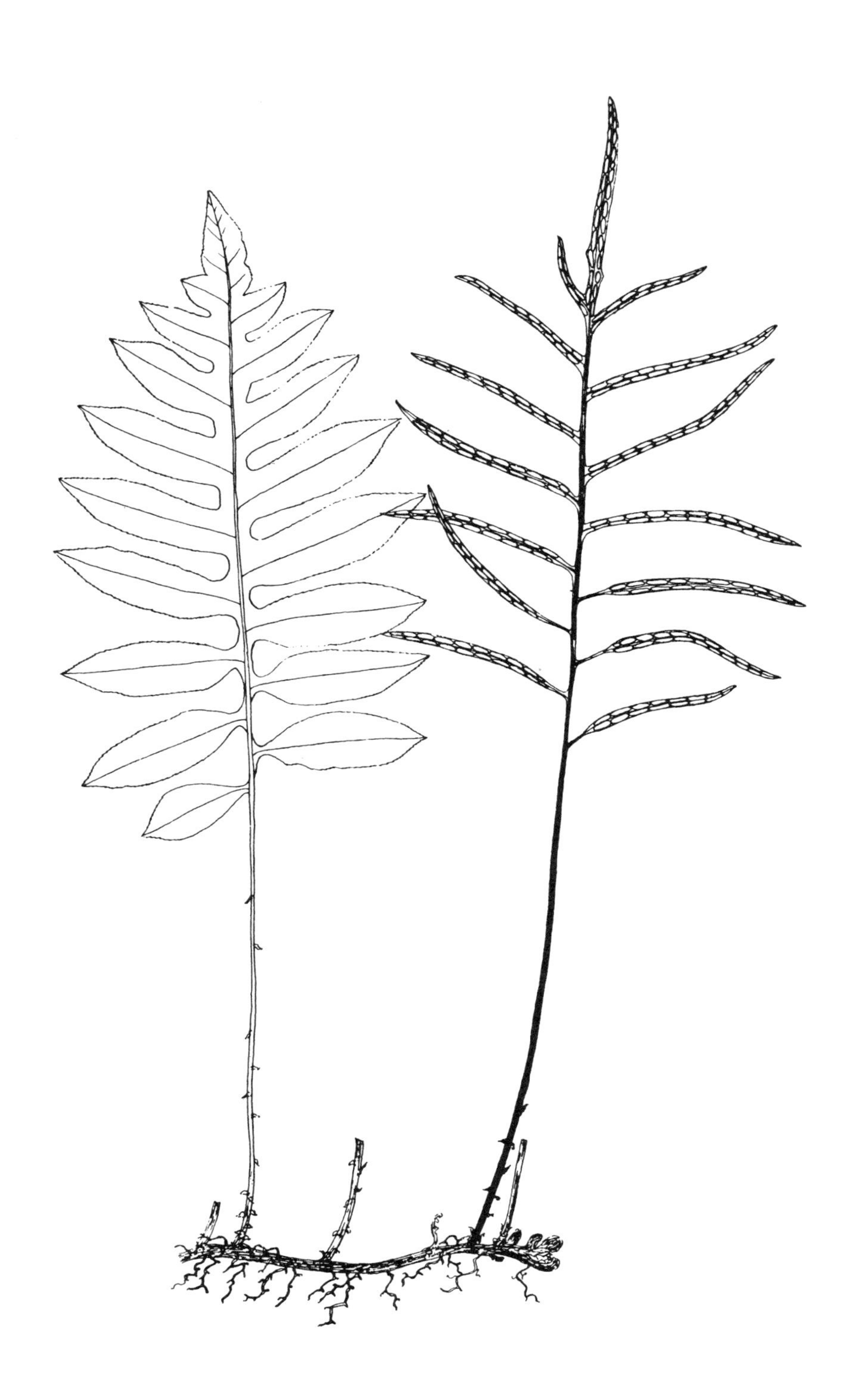

Southern Lady Fern

Athyrium filix-femina (L.) Roth
var. *asplenioides* (Michx.) Farwell

Name: In 1753 Linnaeus described the Eurasian form of the lady fern as *Polypodium filix-femina.* Roth established the genus *Athyrium* in 1799 and included this fern in it. The genus name, meaning "without a door," refers to the tardily opening indusium. In 1803 Michaux named the lady fern of the southeastern United States *Nephrodium asplenioides,* which Eaton transferred to *Athyrium* in 1817. Although some botanists regard it as a distinct species, *A. asplenioides* is generally considered a variety of the more inclusive taxon as determined by Farwell in 1923. The name *filix-femina* means literally "lady fern," and *asplenioides,* "like *asplenium.*"

Rootstock: Short-creeping, ascending or erect. Scaly.

Fronds: 30 to 120 cm long. Clustered, arching, deciduous.

Stipe: 10 to 30 cm long. Yellowish green or reddish. Both colors may be found near each other. Nearly glabrous, except for a few scattered, chaffy scales.

Rachis: Yellowish green or reddish. Smooth. Flat or slightly grooved in front.

Blade: 20 to 90 cm long and 15 to 30 cm wide. Ovate or oblanceolate, broadest near base. Bipinnate to tripinnate, finely cut and delicate.

Pinnae: To 15 cm long and 3 cm wide. Alternate, distant below. Lanceolate with narrow, pointed tips. Lowest one or two pairs generally shorter than those above. Pinnules deep-cut and toothed, blunt-tipped.

Sori: Numerous. Elongate, straight, slightly curved or kidney-shaped. In two rows midway between midvein and margin. Indusium present.

Habitat: Swampy woods, moist, wooded hillsides, roadside banks, and along streams. In most soils except extremely acidic ones.

Range: There are four varieties of lady ferns in North America. The southern lady fern, var. *asplenioides,* ranges from Texas to northern Florida and northward to Massachusetts and Missouri. It is the only variety found in Georgia, where it is one of the most abundant ferns. It is found throughout the state, except rarely in the sandy pineland of southeastern Georgia.

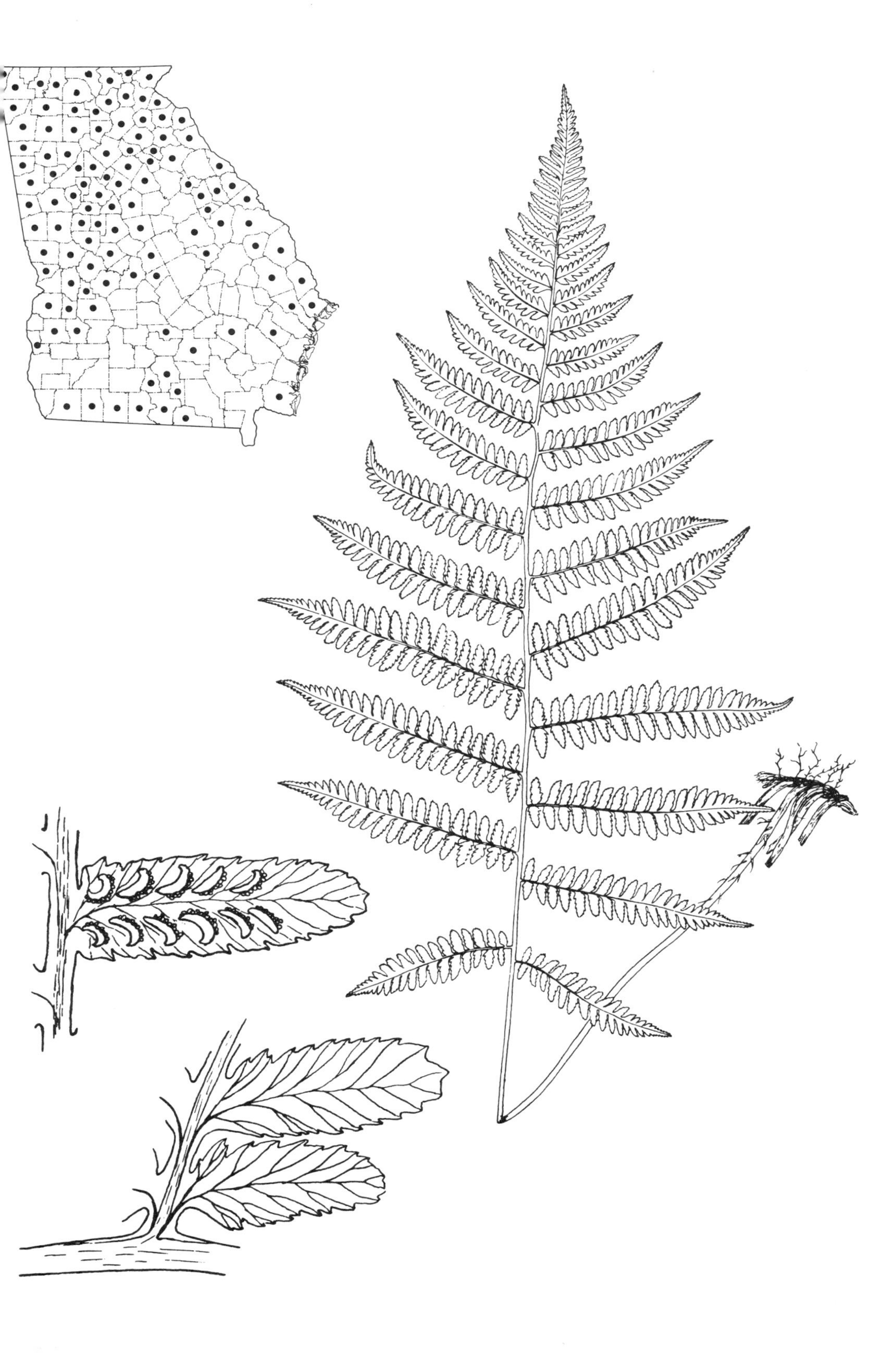

Silvery Glade Fern

Athyrium thelypterioides (Michx.) Desv.

Name: Michaux found this plant in Virginia or North Carolina and in 1803 named it *Asplenium thelypterioides,* the epithet referring to another genus of ferns. In 1827 Desvaux transferred it to *Athyrium.* The common name is given because of the silvery appearance of the frond backs.

Rootstock: Slender, creeping.

Fronds: 50 to 100 cm long. In short rows. Brittle and easily broken. Light green. Deciduous.

Stipe: 15 to 30 cm long. Green above, dark below. Covered with narrow scales and fine hairs.

Rachis: Green. Covered with scales and hairs.

Blade: 35 to 70 cm long and 15 to 25 cm wide. Pinnate-pinnatifid. Decidedly narrowed at base. Finely scaled beneath.

Pinnae: Oblong or linear-oblong with long, acuminate apex. Deeply cut into blunt-ended segments.

Sori: Long, obliquely placed in rows on each side of pinna midvein. Indusia laterally attached.

Habitat: Moist, rich woods and along wet stream banks. Calcareous or noncalcareous soils fairly rich in humus.

Range: Asia and the eastern United States and Canada southward to Louisiana and northern Georgia. In Georgia, in several northern counties southward to Fulton County.

Remarks: There are two forms of *A. thelypterioides.* On one (as found in Georgia) the pinnules are nearly smooth-margined and round-tipped. The other, more northern form has the segments toothed and slightly pointed at the tip.

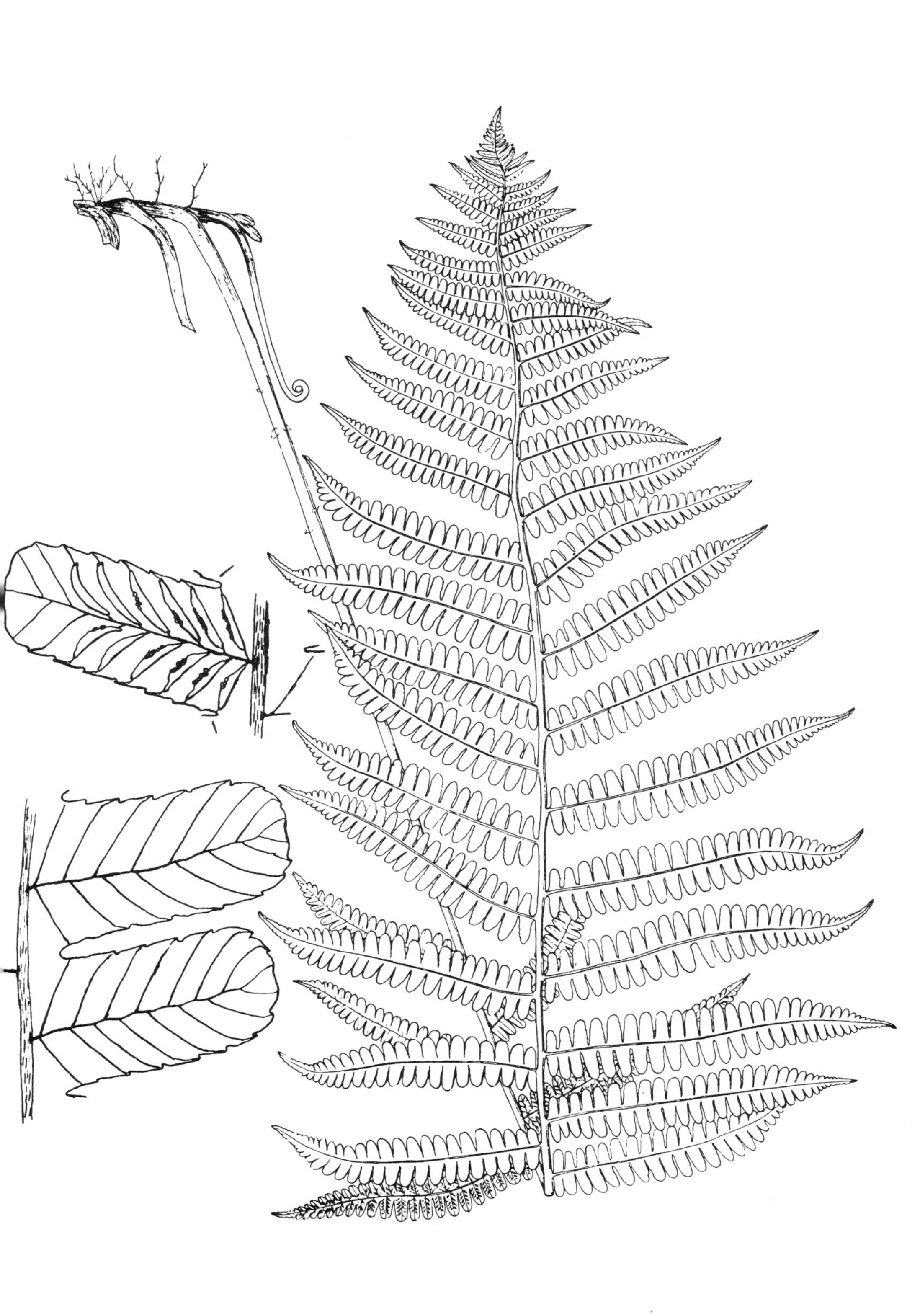

Glade Fern

Athyrium pycnocarpon (Spreng.) Tidestr.

Name: André Michaux found and named this *Asplenium angustifolium* in 1803. As this name was already in use, in 1804 Sprengel proposed *Asplenium pycnocarpon,* which Tidestrom transferred to *Athyrium* in 1906. Michaux's epithet means "narrow-leaved," a common name often used for this fern. *Pycnocarpon,* meaning "crowded fruit," refers to the closely crowded sori on mature plants.

Rootstock: Short-creeping, stout, scaly.

Fronds: 30 to 75 cm long. Clustered. Dimorphic. Fertile longer and narrower than sterile. Easily injured by frost.

Stipe: 15 to 30 cm. Green, darker at base. Mostly naked.

Rachis: Green. Naked or slightly hairy.

Blade: 15 to 45 cm long and 10 to 20 cm wide. Lanceolate with long, acuminate tip and slightly tapering base. Once pinnate.

Pinnae: Mostly alternate. Linear-oblong with long, pointed tips and rounded base. Margins wavy, entire. Fertile narrower, straighter, and more widely spaced than sterile.

Sori: Numerous. Long, narrow, straight or slightly curved. In two rows along upward sloping veins. Prominent indusia laterally attached.

Habitat: Moist, rich woods and shaded ravines. Neutral or mostly calcareous soils.

Range: Eastern United States and Canada southward to Louisiana and northern Georgia. In Georgia, only in a few extreme northern counties.

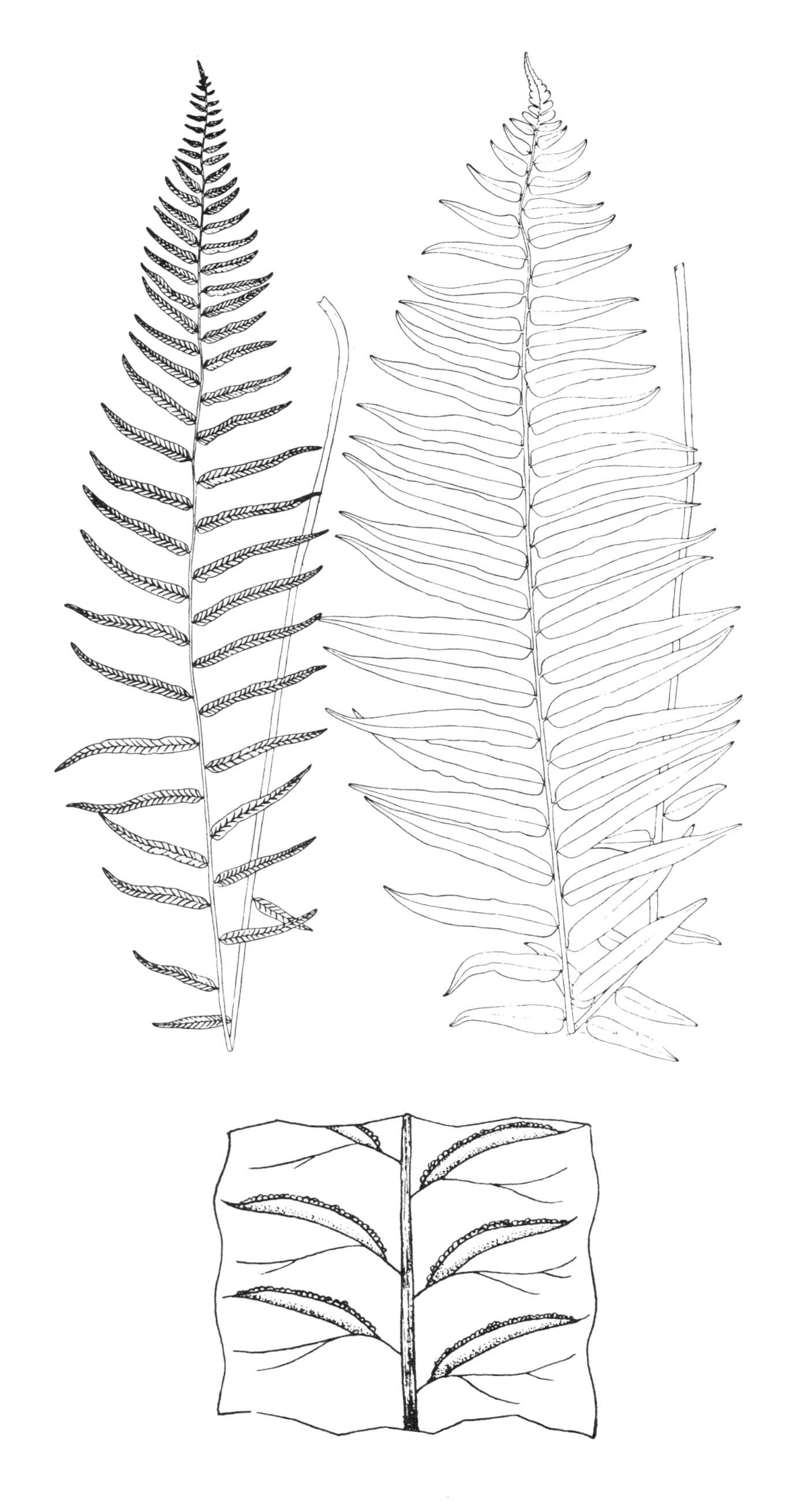

Japanese Twin-Sorus Fern

Diplazium japonicum (Thun.) Bedd.

Name: In 1801 Swartz established the genus *Diplazium,* from the Greek for "doubled," because of the linear grouping of the sori in pairs. In 1784 Thunberg named this fern *Asplenium japonicum,* and in 1876 Beddome transferred it to *Diplazium.* The epithet means "of Japan."

Rootstock: Slender, short-creeping.

Fronds: 30 to 60 cm long. In rows. Dark green above, lighter below.

Stipe: 15 to 30 cm long. Light brown with long, brown scales. Grooved above.

Rachis: Light brown, scaly, grooved.

Blade: 15 to 30 cm long and 10 to 20 cm wide. Oblong-triangular with long, tapering tip. Has short, twisted hairs and narrow scales.

Pinnae: Mostly opposite, widely spaced. Oblong with long, tapering tips. Pinnules somewhat rounded.

Sori: Elongate along veins, occasionally back to back. Indusia prominent, laterally attached.

Habitat: Moist woods along stream banks. Disturbed moist soils.

Range: Common in Asia. Rarely escaped in Florida, Alabama, and Georgia.

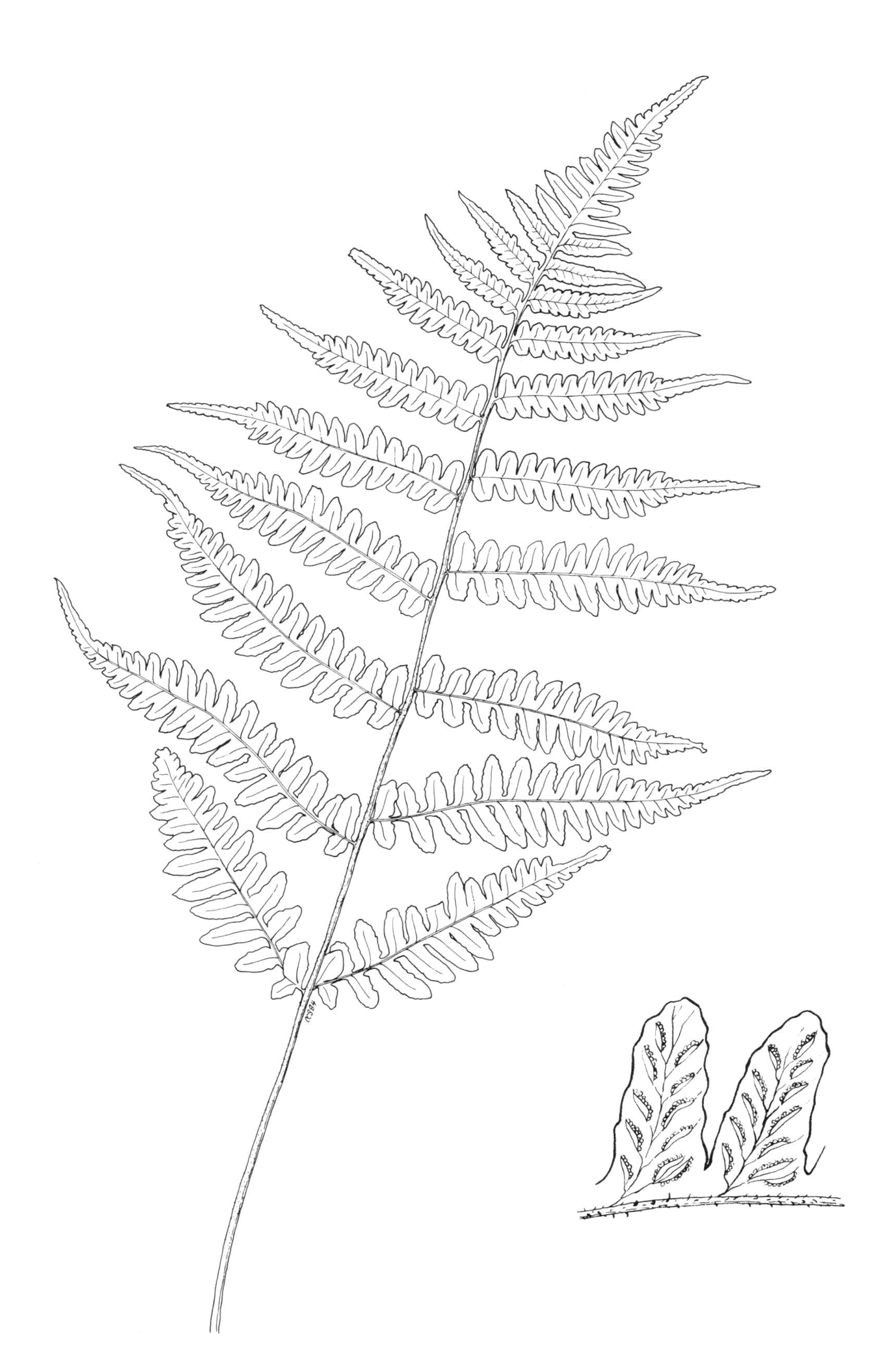

Brittle Fern

Cystopteris protrusa (Weath.) Blasdell

Name: In 1806 Bernhardi proposed the genus *Cystopteris,* meaning "bladder fern," in reference to the bladderlike indusium. For many years this taxon was considered the same species as *Cystopteris fragilis* (L.) Bernhardi. In 1935, however, Weatherby recognized it as *C. fragilis* var. *protrusa,* and in 1960 Blasdell raised it to species rank. *Protrusa,* meaning "protruding," refers to the rhizome extending beyond the stipe.

Rootstock: Wide-creeping. Rhizome tips extend beyond the stipe bases.

Fronds: 25 to 50 cm long. Delicate, lacy, produced in loose clusters. Dark green above, lighter green below. Appear in early spring; may disappear during summer drought and reappear in fall.

Stipe: 12 to 25 cm long. Light green or tan, darkened at base. Glabrous, crisp and brittle. Grooved on upper surface.

Rachis: Green, glabrous. Basal half grooved on upper surface, apical half flat.

Blade: 13 to 25 cm long and 4 to 12 cm wide. Lanceolate or oval, slightly narrowed at base. Apex long, acuminate. Bipinnate-pinnatifid.

Pinnae: Alternate. Ovate to lanceolate with short petioles. Lowest pairs more distant than those above. Lobed or toothed. Free veins end in notches between teeth.

Sori: Roundish, astride veins, submarginal. Indusia hoodlike, laterally attached, conspicuous at first but shrivel away in mature sori.

Habitat: Moist woodlands in humus-rich soils, rocky crevices, along streams, and on sheltered moist ledges. Calcareous or circumneutral soil.

Range: New York to Minnesota and southward, but primarily in the southeastern states, where it is common. In Georgia, in a number of northern counties southward to Fulton County and in Grady County in southern Georgia.

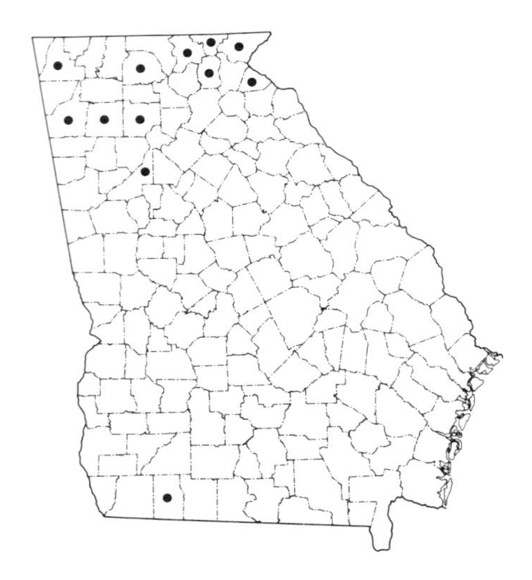

Bulblet Bladder Fern

Cystopteris bulbifera (L.) Bernh.

Name: In 1753 Linnaeus named this fern *Polypodium bulbiferum* from a North American plant. Bernhardi transferred it in 1806 to his newly proposed genus *Cystopteris.* The epithet refers to the "bulblet-bearing" fronds.

Rootstock: Short-creeping, brown-scaly.

Fronds: 25 to 60 cm long. Clustered, lax, rarely standing erect. Delicate, deciduous, pale green.

Stipe: 5 to 15 cm long. Slender, glabrous, pinkish with dark base.

Rachis: Straw-colored, glabrous.

Blade: 20 to 45 cm long and 8 to 15 cm broad at base. Long, narrowly triangular. Widest at or near base. Tapers gradually and uniformly to very long apex. Bipinnate-pinnatifid.

Pinnae: Over 20 pairs, nearly opposite, somewhat ascending. Sessile or with very short petioles. Lanceolate. Pinnules oblong, toothed.

Sori: Orbicular. Astride veins. Indusia hood-shaped with truncated apex.

Habitat: Usually on or among limestone rocks. Also on talus slopes and shaded ledges.

Range: Common in temperate central and eastern North America. Ranges from the Atlantic coastline states to the midwestern states and southward from Canada to Georgia, Alabama, and Mississippi. In Georgia, restricted to a few sites in the mountains of the Ridge and Valley Province.

Remarks: Asexual reproduction by small, green, fleshy bulblets borne on underside of fronds near the axils of pinnae or pinnules, which, after dropping off, may grow and produce new plants.

Tennessee Bulblet Fern

Cystopteris tennesseensis Shaver

Name: Jesse M. Shaver discovered this fern in Tennessee and in 1950 named it after that state.

Rootstock: Short and slender. Usually horizontal or slightly inclined. Covered with old stipe bases.

Fronds: 22 to 45 cm long. Dark green above, lighter green below.

Stipe: 10 to 15 cm long. Blackish brown at base, green or straw-colored near base of blade. Scales usually limited to basal portion.

Rachis: Green to straw-colored. Glabrous.

Blade: 12 to 30 cm long and 5 to 15 cm wide. Narrowly deltoid to lanceolate. Broadest at or near base, with acuminate tip. Glabrous. Bipinnate.

Pinnae: Basal ovate-deltoid or lanceolate with acute apex. Upper pinnae lanceolate with shorter stems. Broadly attached segments replace pinnae at apex of blade. Pinnules are mostly lanceolate with short stems and toothed lobes.

Sori: Orbicular. Astride veins. Indusia small, attached under sori. With time, sori enlarge and overgrow indusia. Bulblets occasionally grow in or near axils of pinnae.

Habitat: Small, wet, soil-filled crevices of damp, north-facing limestone bluffs.

Range: Tennessee, the Ozarks, Kansas, and northern Georgia.

Remarks: *Cystopteris tennesseensis* is a fertile hybrid of *C. protrusa* and *C. bulbifera.* As with *C. bulbifera,* the bulblets often reproduce plants asexually.

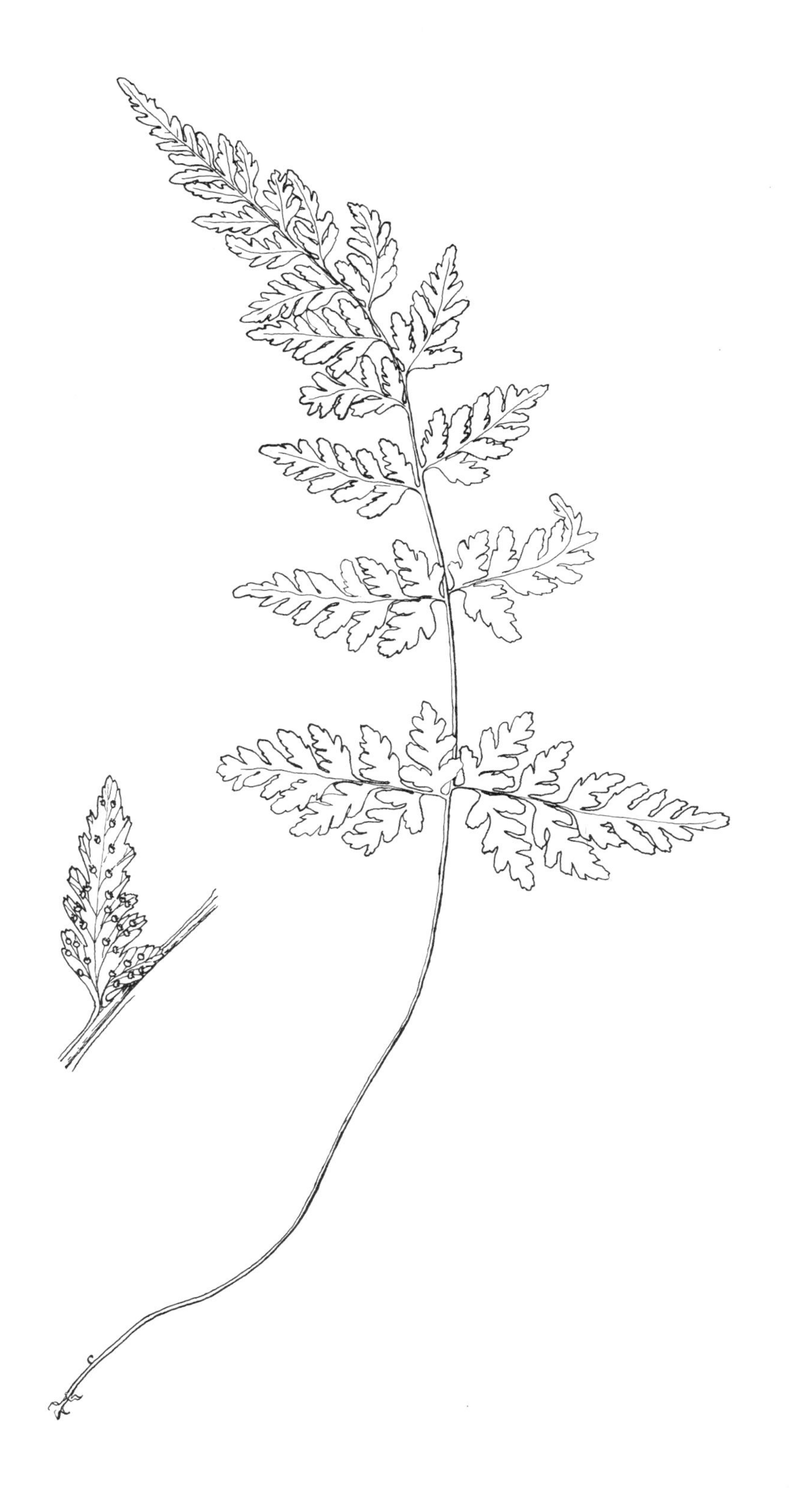

Blunt-Lobed Woodsia

Woodsia obtusa (Spreng.) Torr.

Name: This species was discovered in Pennsylvania and described by Sprengel in 1804 as *Polypodium obtusum.* Torrey in 1840 transferred it to *Woodsia,* a genus established by Robert Brown in 1813 and named in honor of Joseph Woods. The epithet refers to the blunt lobes of the pinnae. It is also commonly called "cliff fern."

Rootstock: Short-creeping. Scales tan, later with dark stripes.

Fronds: 20 to 50 cm long. Clustered, light green, delicate. Fertile, deciduous. Sterile, shorter, broader, evergreen.

Stipe: 10 to 20 cm long. Green, becoming straw-colored with dark base. Conspicuous, light brown scales.

Rachis: Green or straw-colored. Slightly scaly.

Blade: 10 to 30 cm long and 6 to 12 cm broad. Pinnate-pinnatifid. Oblong. Widest near middle.

Pinnae: Distant, almost opposite. Deltoid to lanceolate. Pinnules oblong-obtuse, winged at axis, with blunt tips. Margins crenate.

Sori: Round, marginal. Indusia cup-like, splitting into segments and giving starlike appearance.

Habitat: Shaded, wooded hillsides and woodlands. Often on rocky ledges and in crevices.

Range: Common throughout eastern North America from Canada to Florida and westward to Oklahoma and Kansas. Found throughout Georgia, especially in the northern half.

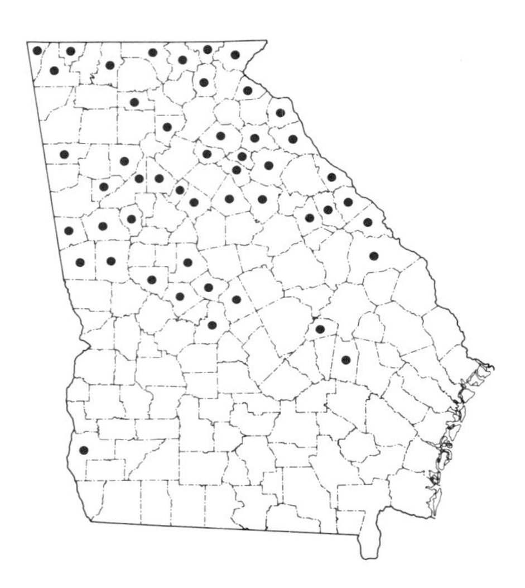

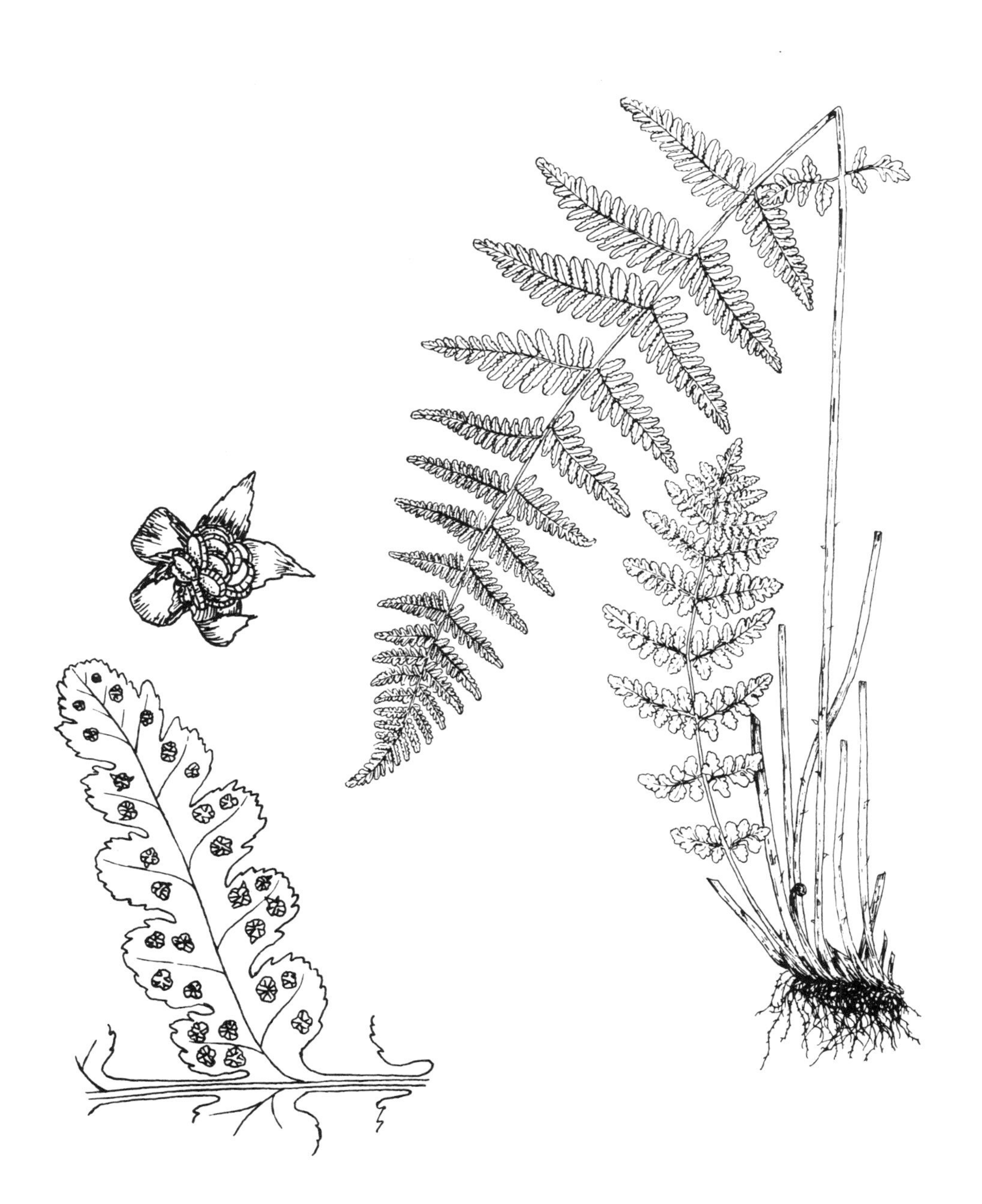

Rocky Mountain Woodsia

Woodsia scopulina D. C. Eaton

Name: D. C. Eaton described this fern in 1865. The epithet means "of rocky places."

Rootstock: Stout, ascending, with narrow, tan scales.

Fronds: 10 to 40 cm long. Tufted. Light green. Delicate, deciduous.

Stipe: 4 to 12 cm long. Shining chestnut brown. Tan scales, especially at base, and white hairs.

Rachis: Straw-colored with white hairs.

Blade: 6 to 28 cm long and 3 to 8 cm wide. Oblong to linear. Acuminate at apex and narrowed at base. White hairs and tiny glands. Bipinnate to bipinnate-pinnatifid.

Pinnae: Alternate, well spaced. Deltoid to lanceolate. Pinnules oblong, slightly winged at axis. Margins crenate.

Sori: Round, marginal. Indusia cut into ribbonlike segments.

Habitat: Cliffs and ledges of rock. Soil mostly subacidic.

Range: Frequent in the mountainous areas of northern and western North America. Also in the Ozarks and southern Appalachians.

Sensitive Fern

Onoclea sensibilis L.

Name: *Onoclea* is an ancient plant name that Linnaeus applied to this fern in 1753. The fern is "sensitive" only in that the sterile fronds are easily killed by frost. "Bead fern," referring to the beadlike fertile pinnules, is actually more appropriate.

Rootstock: Slender, green, long-creeping near soil surface.

Fronds: Dimorphic. Distant, produced singly. Sterile, 35 to 75 cm long. Coarse, deciduous, arise in spring or early summer. Fertile, 25 to 40 cm long. Arise in midsummer, persist through winter. At first dark green, but soon become brown.

Stipe: Sterile, 20 to 40 cm long. Green to straw-colored with dark base. Fertile, 15 to 25 cm long. Cylindrical, upright, woody and rigid, dark brown.

Rachis (of sterile blade): Green or straw-colored near base, brownish near apex. Grooved on upper side except near apex. Winged, with wings broadest near apex.

Sterile Blade: 15 to 35 cm long and broad. Dark green above, lighter below. Deltoid-ovate with prominent, netted veins. Pinnatifid. Cut into about 12 pairs of nearly opposite, narrowly oblong segments. Basal pair with short stalks, others winged at axis. Margins wavy to shallowly lobed, entire.

Fertile Blade: 10 to 15 cm long. Pinnate, with erect pinnae 2 to 7 cm long, each bearing two rows of beadlike structures, which are leaf segments wrapped around a group of sori.

Sori: Contained in beadlike structures grouped at upper end of fertile stalk. Spores released the following spring.

Habitat: Swampy woodlands, muddy ditches, wet soil in meadows and river swamps. Thrives in often-flooded situations. Soil slightly acidic or circumneutral.

Range: Common throughout most of eastern temperate North America. In Georgia, found throughout the state, but less frequently in the southeastern Coastal Plain.

Remarks: *Onoclea sensibilis* forma *obtusilobata* Gilbert is an occasional intermediate between sterile and fertile fronds. It resembles the sterile but is usually smaller. The pinnae are not winged and are cut into more and deeper segments. The veins mostly end free without forming areoles.

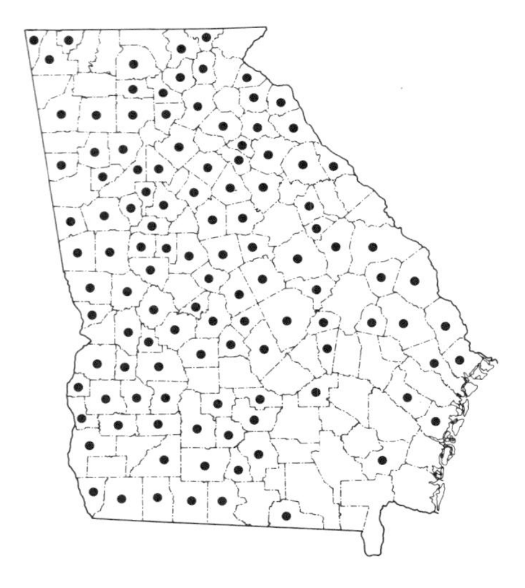

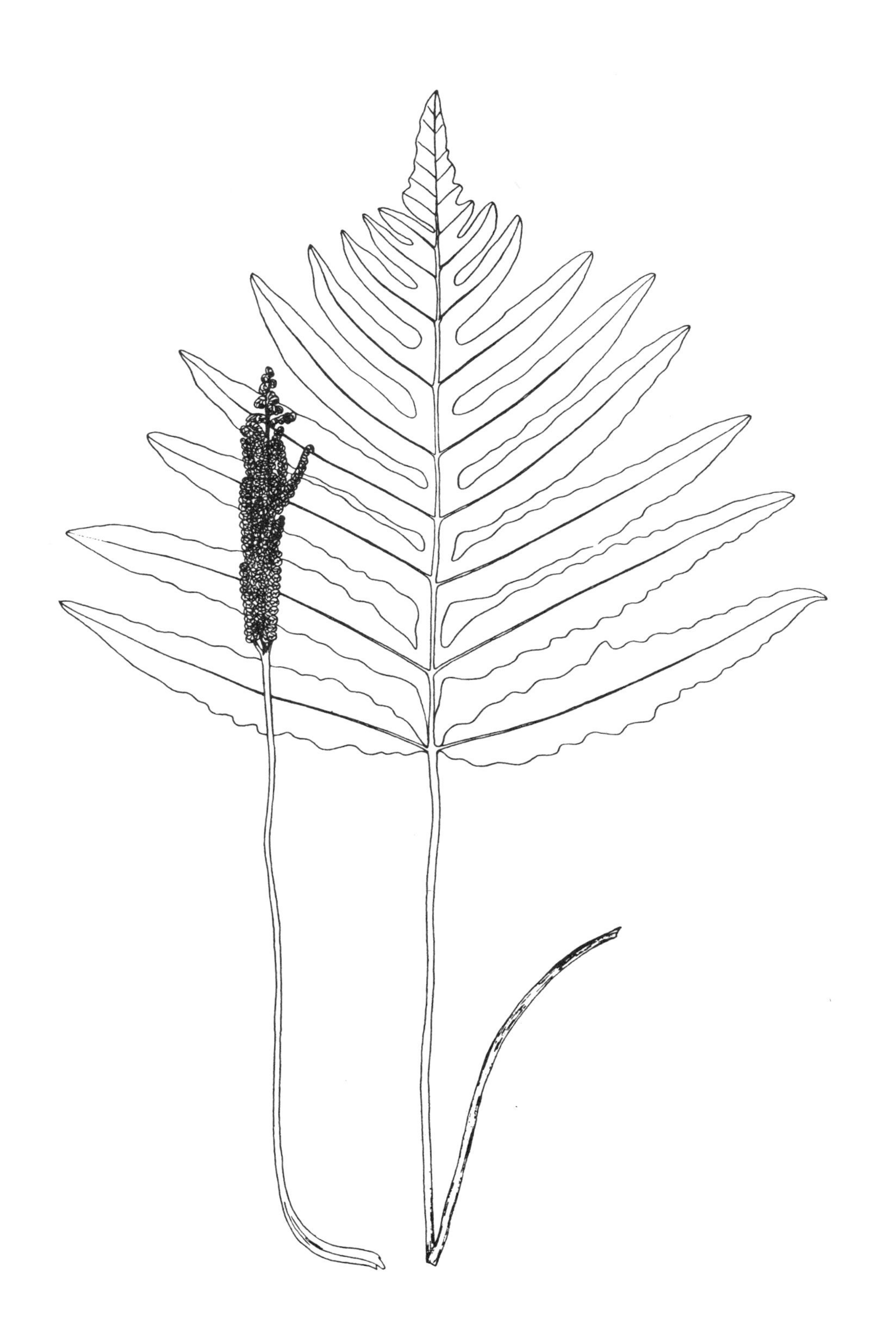

Mariana Maiden Fern

Thelypteris torresiana (Gaud.) Alston

Name: The genus name *Thelypteris,* meaning "a female fern," was first used by Linnaeus as a specific epithet. In 1762 Schmidel used it as a generic designation. Since "lady fern" applies to a separate genus and species, the common term "maiden fern" is used for the genus. Some botanists divide the group into several genera (e.g., *Cyclosorus, Phegopteris,* etc.), although the most common practice is to consider these as sections of *Thelypteris.* For many years botanists considered most of the ferns to be included in *Dryopteris.* Gaudichaud named this fern *Polystichum torresianum* in 1828 in honor of D. L. de Torres, and in 1960 Alston placed it in *Thelypteris.* The common name refers to the fact that it is a native of the Mariana Islands.

Rootstock: Short-creeping, with long, brown scales.

Fronds: 60 to 125 cm long. Clustered, spreading.

Stipe: 25 to 50 cm long. Stout, light-colored.

Rachis: Whitish.

Blade: 35 to 75 cm long and 25 to 50 cm broad at base. Broadly triangular. Bipinnate-pinnatifid. Covered with needlelike hairs on both surfaces.

Pinnae: Alternate, widespread. Lanceolate. Broadest at base with long, tapering apex. Pinnules sharp-toothed.

Sori: Numerous, circular. In rows along each side of pinnule midvein. Indusia short-lived or absent.

Habitat: Moist woods, along stream banks and under highway bridges.

Range: First found in the United States in Florida; since then has spread to other southeastern states. In recent years has been found more and more in southern Georgia and up to Jasper and Putnam counties, indicating that it is probably continuing to spread.

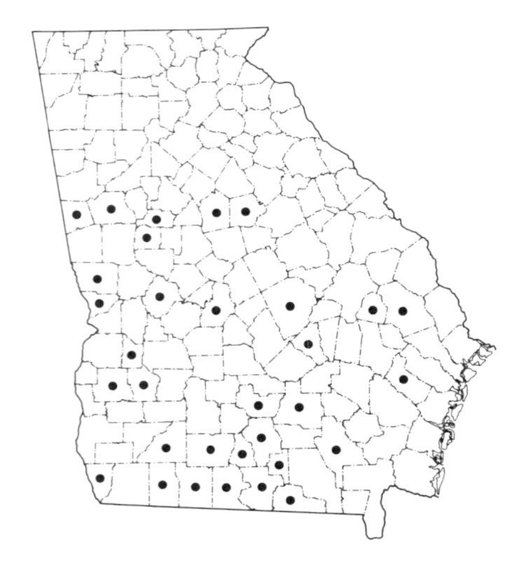

Broad Beech Fern

Thelypteris hexagonoptera (Michx.) Weath.

Name: Michaux found this fern in the southeastern United States in 1803 and described it as *Polypodium hexagonopterum.* Weatherby transferred it to *Thelypteris* in 1919. The specific name means "6-angled wing," in allusion to the broad-angled wing of the rachis.

Rootstock: Slender. Long-creeping.

Fronds: 35 to 70 cm long. In rows. Delicate and easily broken, neither persistent nor evergreen. Dull green. Fertile, relatively few and appearing late.

Stipe: 20 to 40 cm long, as long as or longer than blade. Light green or straw-colored with dark base.

Rachis: Green. Winged throughout.

Blade: 15 to 30 cm long and 15 to 35 cm wide. Bipinnatifid. Broadly triangular, broadest at base, usually broader than long. Long, pointed apex. Lower surface with fine hairs. Segments lanceolate with long, pointed apex and often tapering at base. Connected to each other by wings on rachis. Lowermost longest and inclined forward. Ultimate segments oblong with rounded tips and winged at midvein.

Sori: Small, round, marginal. Indusia lacking.

Habitat: Humus-rich, well-drained woodlands and cool, shaded slopes of ravines and rocky gorges.

Range: Eastern North America from southern Canada southward to eastern Texas and northernmost Florida. Common throughout northern Georgia, ranging sparingly southward to the Florida line in southwestern Georgia.

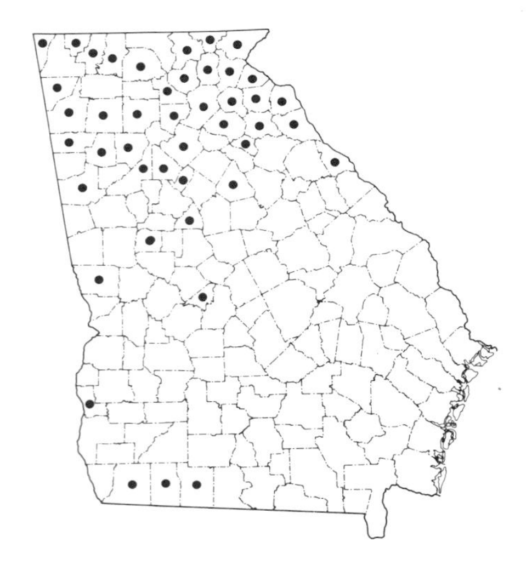

Downy Maiden Fern

Thelypteris dentata (Forssk.) E. St. John

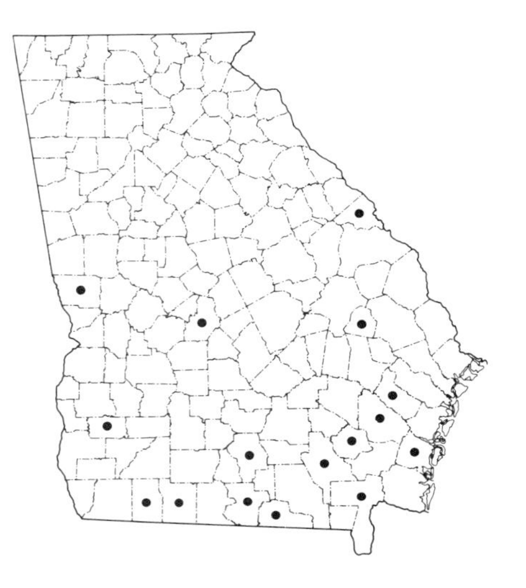

Name: The downy maiden fern, or tapering tri-vein fern, was named *Polypodium dentatum* by Forsskal in 1775. It was placed in *Thelypteris* by E. P. St. John in 1936. The epithet means "toothed."

Rootstock: Stout. Short-creeping. Slender, brown scales.

Fronds: 40 to 100 cm long. Tufted. Dark green. Sterile smaller and with wider pinnae than fertile.

Stipe: 20 to 40 cm long. Purplish with scattered short hairs.

Rachis: Purplish. Hairy.

Blade: 20 to 60 cm long and 15 to 25 cm wide. Pinnate-pinnatifid. Lanceolate with long, tapering tip.

Pinnae: 1.5 to 2.5 cm wide. Oblong, abruptly contracted to tapering tip. Shallowly cut into broad, blunt lobes. Lowermost shortest and turned downward. Short hairs on both upper and lower surfaces. Basal veins from adjoining lobes fuse near the pinna midvein to form a single vein running to sinus.

Sori: Small, medial. Indusia kidney-shaped.

Habitat: Damp woods, roadsides, pastures, under bridges.

Range: Native of tropics, rather recently introduced to New World. Found in the Gulf states. In Georgia, primarily in lower sections of the Coastal Plain.

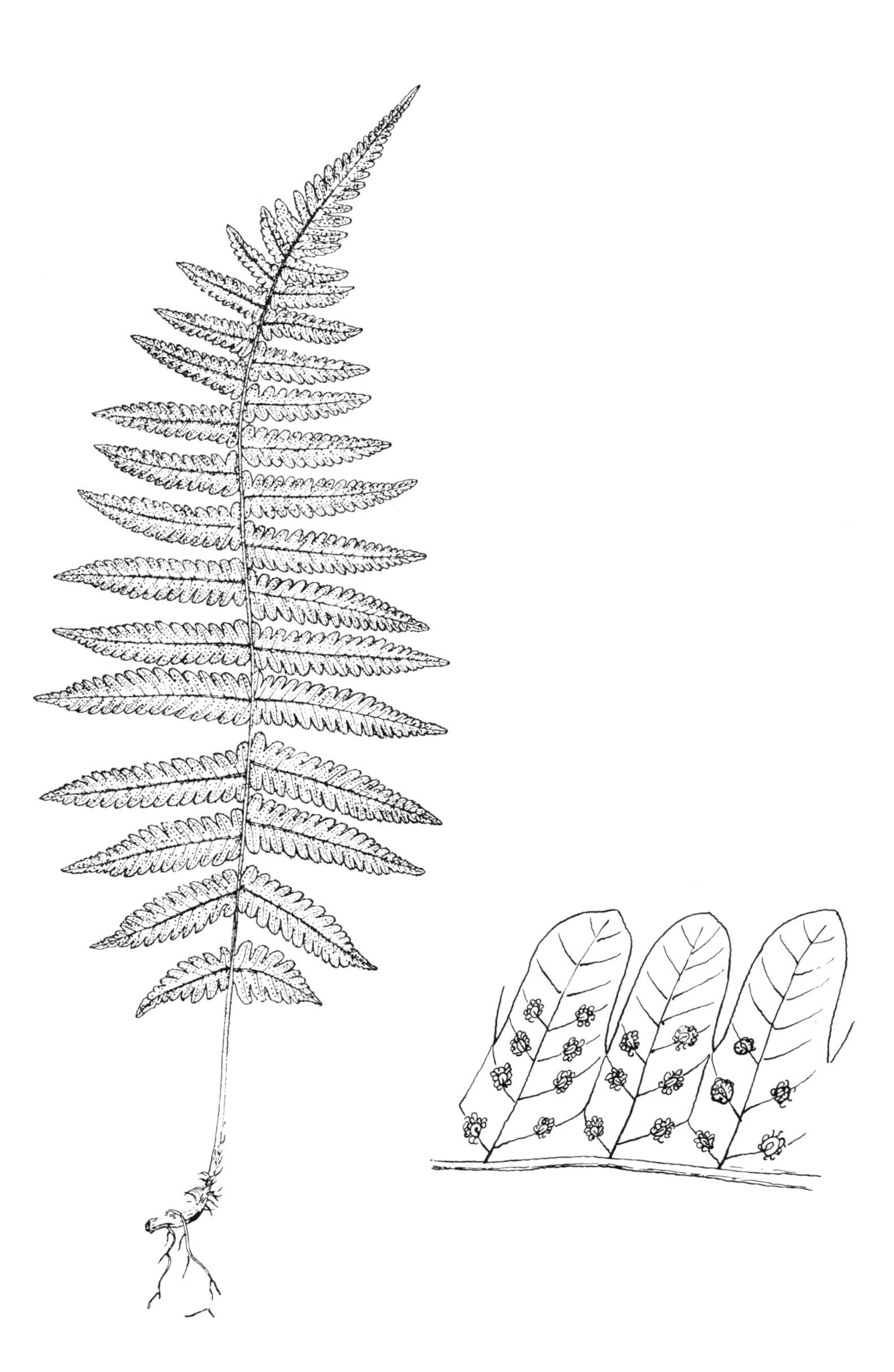

Variable Maiden Fern

Thelypteris hispidula (Decne.) Reed var. *versicolor* (R. St. John) A. R. Smith

Name: For some years this plant was known as *Thelypteris quadrangularis* (Fée) Schelpe. In 1968, however, Reed suggested that it was the same as the Old World plant *Aspidium hispidulum,* named by Decaisne in 1835. In 1936 Robert St. John called Florida specimens *T. versicolor* because of the color variation of the blade, but in 1971 A. R. Smith determined this fern to be merely a variety of the more inclusive species. *Hispidula* means "with short hairs."

Rootstock: Short-creeping to erect. Stout, with long, narrow scales.

Fronds: 40 to 80 cm long. Clustered. Fertile and sterile alike.

Stipe: 10 to 30 cm long. Straw-colored. Hairy.

Blade: 30 to 50 cm long and 10 to 30 cm wide. Pinnate-pinnatifid. Ovate-lanceolate, evenly tapered to short, pinnatifid apex. Both surfaces hairy; hairs on lower surface variable in length.

Pinnae: Long-oblong with pointed tips. Alternate, mostly descending. Cut two-thirds or more to midvein. Basal 1 or 2 pairs may be slightly reduced. Veins variable. Basal veins of adjacent pinnae may unite to form a vein going to the sinus, or may run side by side to the sinus without uniting. Both types may be on the same frond.

Sori: Submarginal. Indusia peltate, kidney-shaped.

Habitat: Complete or partial shade. Moist woods, hammocks, limestone sinks, along stream banks.

Range: In the coastal United States from South Carolina to eastern Texas. In Georgia, in south-central and southwestern counties.

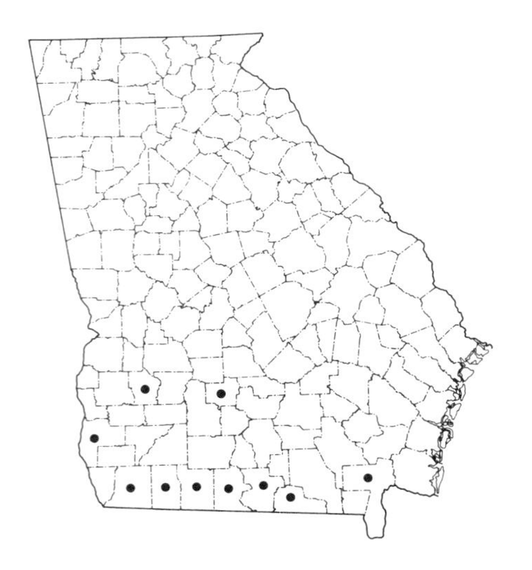

Widespread Maiden Fern

Thelypteris kunthii (Desv.) Morton

Name: This fern has for some years been known as *Dryopteris normalis,* a name Christensen proposed in 1910, or as *Thelypteris normalis,* as named by Moxley in 1920. In 1967 C. V. Morton determined this fern to be the same plant the French taxonomist Desvaux in 1827 had named *Nephrodium kunthii* in honor of the botanist Kunth, hence the current *T. kunthii.*

Rootstock: Stout. Short-creeping.

Fronds: 55 to 110 cm long. Produced in short rows.

Stipe: 25 to 50 cm long. Straw-colored, dark at base.

Rachis: Straw-colored.

Blade: 30 to 60 cm long and 15 to 40 cm wide. Pinnate-pinnatifid. Lanceolate with long, tapering tip.

Pinnae: Long-oblong, with long, tapering tips. Subopposite. Lowermost slightly or not reduced. Veins and midveins very hairy beneath, slightly hairy above. Hairs below all long. Basal veins of adjacent pinna lobes run to sinus without uniting.

Sori: Medial. Indusia peltate, kidney-shaped, hairy.

Habitat: Variety of wet soils, mostly calcareous. Rock crevices, limesinks, swampy woods, stream banks, drainage ditches, and under bridges.

Range: Central and South America and the West Indies. In the Coastal Plain states from South Carolina to Texas. Common in southern Georgia, ranging northward to Putnam County.

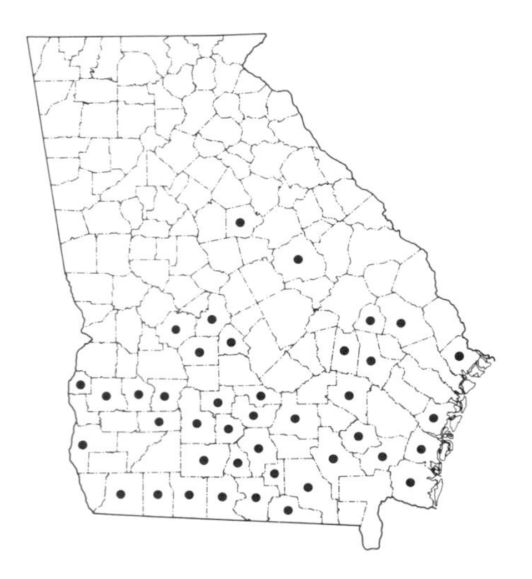

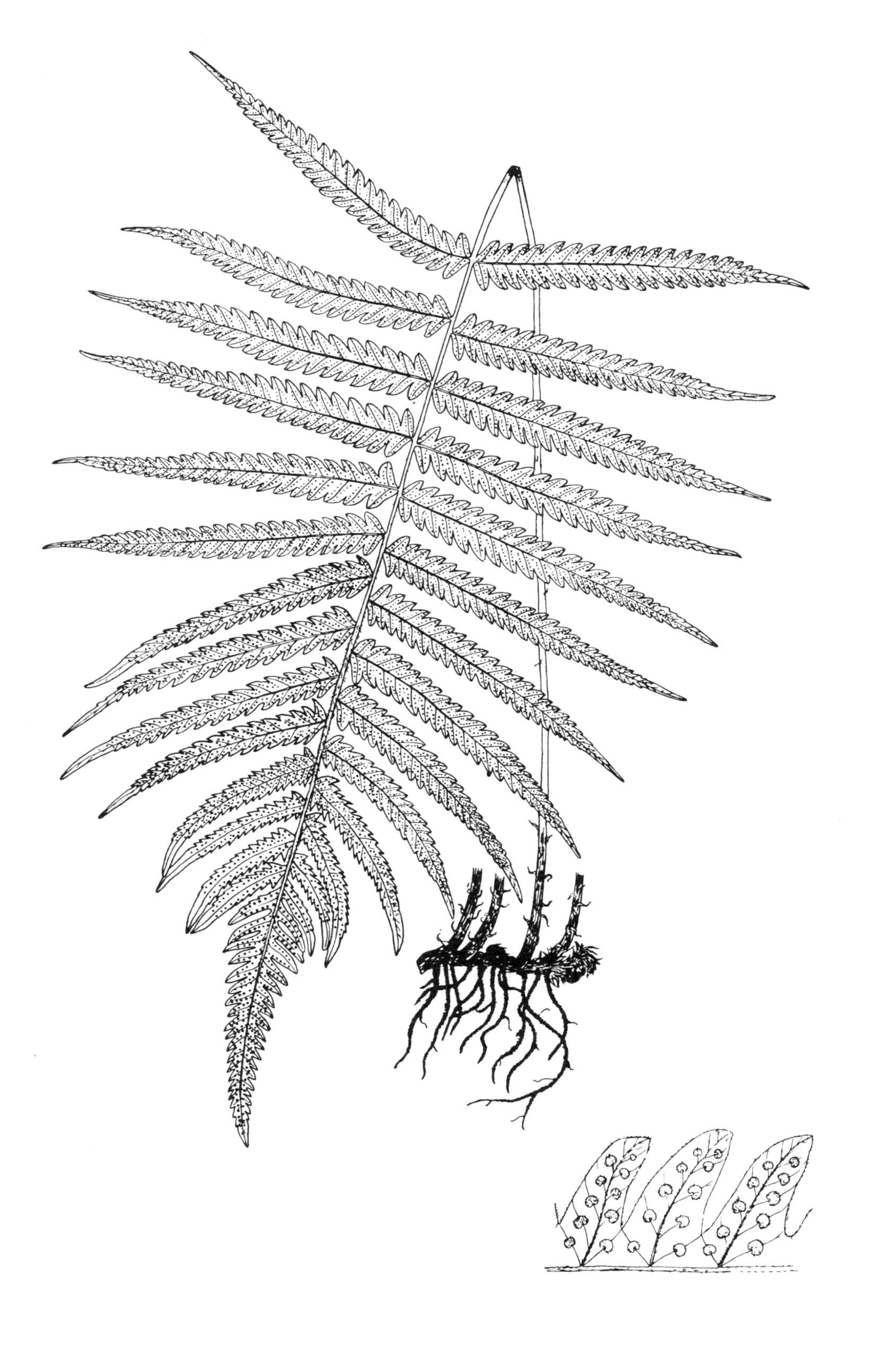

Ovate Maiden Fern

Thelypteris ovata R. St. John

Name: R. M. Harper first discovered this taxon in 1902 along the Ocmulgee River in Pulaski County, and R. St. John named it in 1938. The epithet means "egg-shaped," referring to the general shape of the blade.

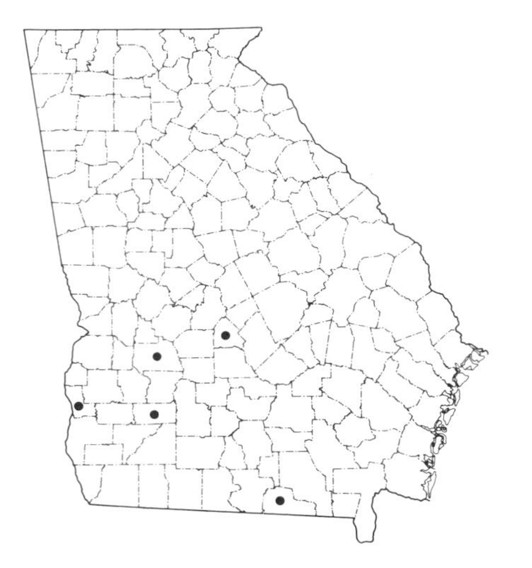

Rootstock: Short- to long-creeping. Slender, branching. Hairy scales.

Fronds: 40 to 110 cm long. In rows.

Stipe: 15 to 35 cm long. Straw-colored, darkened at base.

Rachis: Straw-colored. Sparsely to densely hairy.

Blade: 25 to 75 cm long and 10 to 40 cm wide. Pinnate-pinnatifid. Deltoid-lanceolate, broadly rounded at base, tapers evenly to elongate, pinnatifid apex. Hairs below of moderate density, no hairs above.

Pinnae: Alternate, closely spaced, spreading upward at slight angle. About 1.3 cm wide. Deeply cut, about three-quarters way to midvein. Basal only slightly or not reduced. Basal veins connivent at sinus.

Sori: Few, small, round, submarginal. Hairs on kidney-shaped indusia sparse.

Habitat: Moist woods, hammocks, and limestone sinks.

Range: The Georgia fern, found in the southwestern Upper Coastal Plain, is var. *ovata.* It is found throughout Florida and in a few sites in South Carolina, Alabama, and Georgia.

Marsh Fern

Thelypteris palustris Schott

Name: Linnaeus, from European specimens, described this taxon in 1753 as *Acrostichum thelypteris.* In 1834 Heinrich Schott, an Austrian pteridologist, classified it as *Thelypteris palustris,* meaning "of marshes," where it commonly grows.

Rootstock: Slender, black, creeping and spreading.

Fronds: 45 to 90 cm long. Dull green. Delicate, distant, deciduous.

Stipe: 25 to 50 cm long. Green, with dark base. Smooth.

Rachis: Green, hairy.

Blade: 20 to 40 cm long and 8 to 16 cm wide. Pinnate-pinnatifid. Narrowly oval, with short, tapering apex. Not tapering to base.

Pinnae: Linear-oblong with pointed tips. Mostly alternate, almost sessile. Lowest only slightly reduced. Deeply cut into oblong-obtuse pinnules with margins entire. Veins mostly forked between pinnule midvein and margin. Segments on fertile fronds inrolled, appearing more contracted.

Sori: Round, medial. Partly concealed by curled margins. Indusia kidney-shaped, peltate.

Habitat: Wet, swampy woods, open, grassy marshes, thickets, and swamps.

Range: Europe, Asia, and eastern North America. In Georgia, primarily in the Coastal Plain, where it is widely distributed but nowhere abundant.

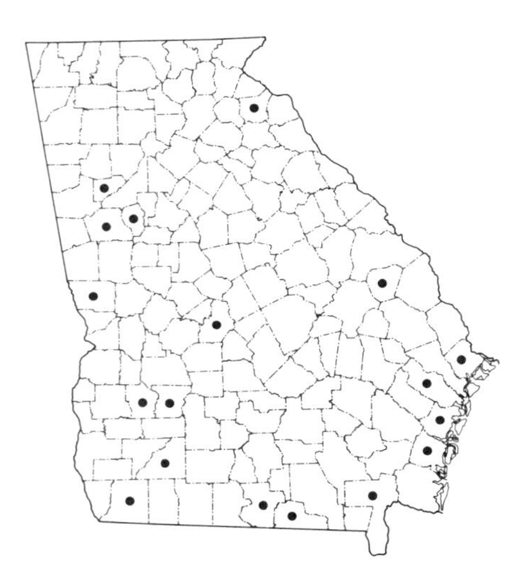

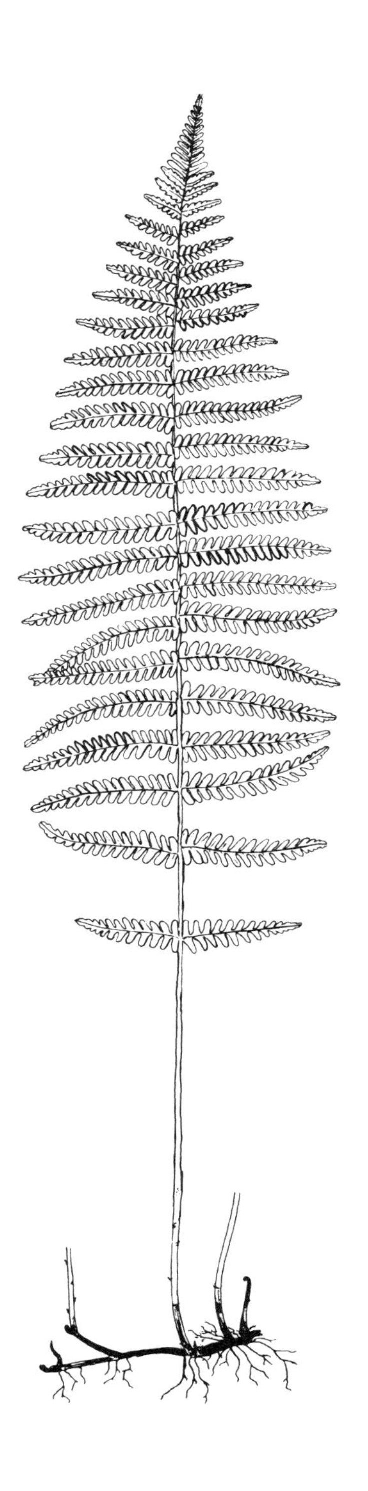

New York Fern

Thelypteris noveboracensis (L.) Nieuwl.

Name: In 1753 Linnaeus placed this taxon in his large genus *Polypodium*, and Nieuwland transferred it in 1910 to *Thelypteris*. Eboracum was the Roman name for the English town of York, and *noveboracensis* means "of New York." As Linnaeus received his specimen from Canada, it has never been explained why he used this epithet.

Rootstock: Slender. Widely creeping.

Fronds: 30 to 60 cm tall. Produced in rows. Light yellow-green. Delicate, deciduous. Fertile slightly larger than sterile.

Stipe: 5 to 10 cm long. Pale green with dark base. Slightly hairy.

Rachis: Pale green. Slightly hairy.

Blade: 25 to 50 cm long and 8 to 15 cm broad. Pinnate-pinnatifid. Elliptical; long, tapering to both ends with lowest pinnae often mere wings and nearly to the ground. Hairy.

Pinnae: Narrowly oblong, tapering to apex. Mostly alternate. Deeply cut into oblong, blunt-tipped, entire pinnules.

Sori: Small, circular, near margins of lobes. Indusia peltate, hairy.

Habitat: Moist or wet thickets or woodlands, banks of streams, and fairly open, swampy places. Neutral or somewhat acidic soil.

Range: Widely distributed in eastern temperate North America southward to Arkansas and northern Georgia. Common in northern Georgia, ranging southward to Harris and Oglethorpe counties.

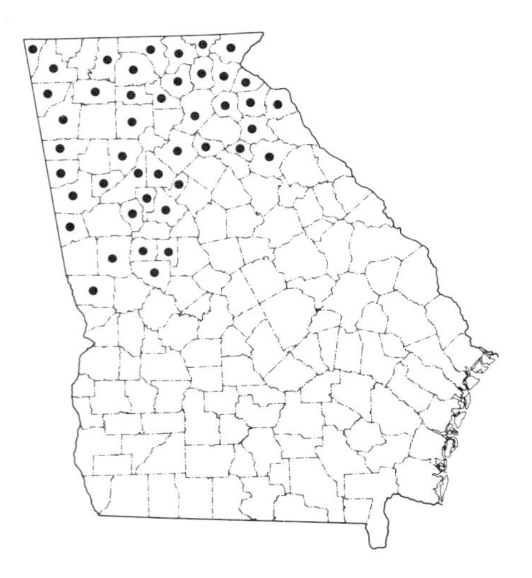

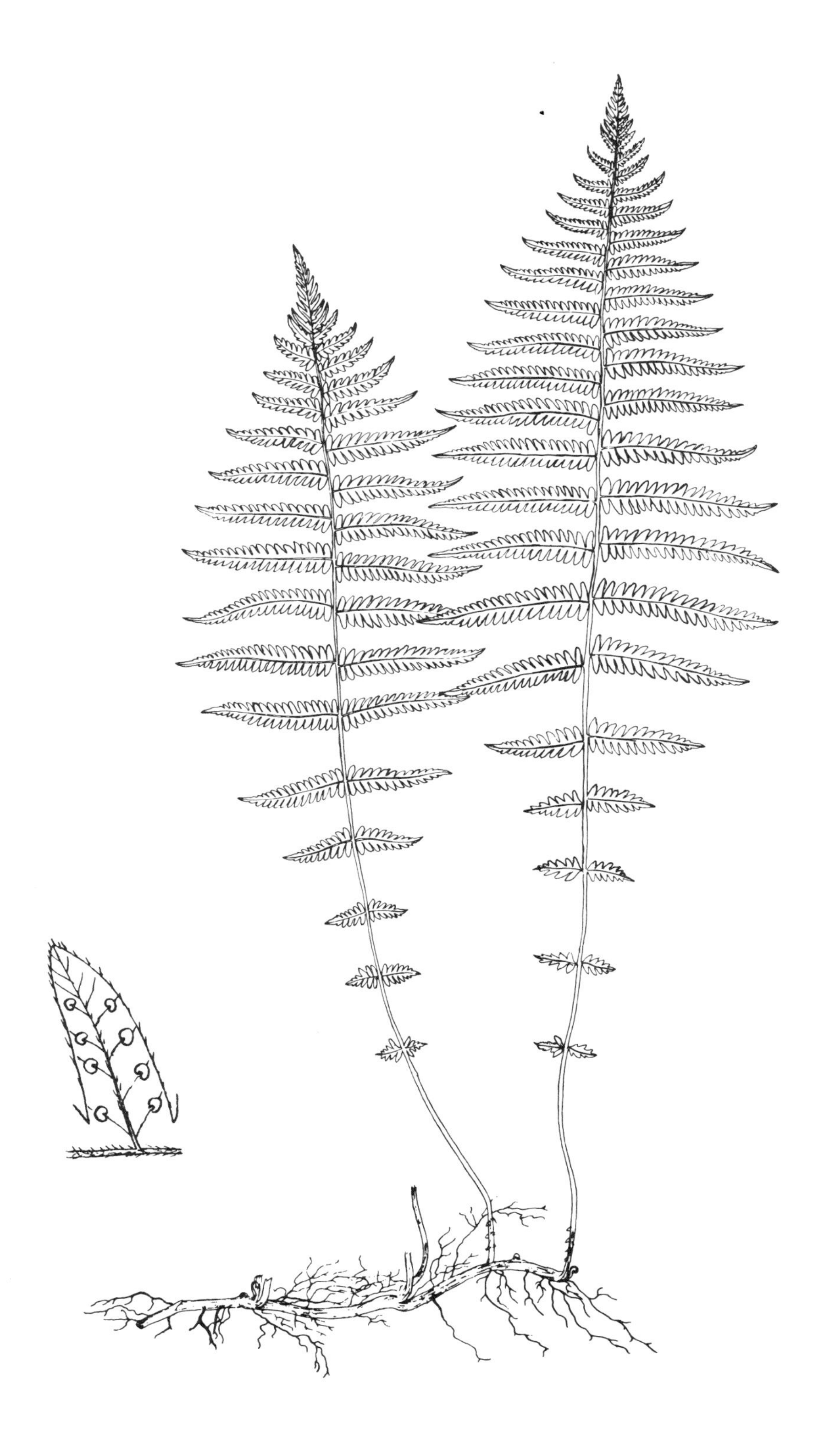

Marginal Wood Fern

Dryopteris marginalis (L.) A. Gray

Name: Michel Adanson, the French taxonomist, established the *Dryopteris* genus in 1763, and Schott redefined it in the presently accepted sense in 1834. The Greek name means "oak fern" for "growing among oaks." In 1753 Linnaeus named *Polypodium marginale* from a North American specimen. Asa Gray transferred it to *Dryopteris* in 1848. The epithet refers to the location of the sori near the margins of the pinnae lobes. Other common names are "marginal shield fern," "evergreen wood fern," and "leather fern."

Rootstock: Short, thick, covered with large, light brown, shining scales.

Fronds: 45 to 60 cm long. Clustered, leathery, evergreen. Fertile appear in spring or early summer and persist until the next year.

Stipe: 15 to 20 cm long. Many pale brown scales, especially at base.

Rachis: Pale, slightly scaly.

Blade: 30 to 40 cm long and 15 to 25 cm wide. Dark green above, much lighter below. Oblong-ovate. Bipinnate to pinnate-pinnatifid.

Pinnae: Alternate, deeply cut. Lanceolate, rapidly taper to a point. Pinnules blunt-tipped with edges shallowly toothed or lobed.

Sori: Marginal. Single or in rows. Large, prominent. Indusia peltate, kidney-shaped, prominent.

Habitat: Rocky, wooded slopes and rich woodlands.

Range: Eastern temperate North America from Canada southward to Oklahoma and northern Alabama and Georgia. Also in British Columbia. In northern Georgia southward to Rockdale County.

Hybrid Wood Fern

Dryopteris × *neo-wherryi* Wagner

Name: This plant was named by Wagner in 1975 in honor of Edgar T. Wherry. The prefix *"neo,"* or new, indicates that previously a *Dryopteris* had been named in honor of Wherry but proved to be an invalid species.

Rootstock: Stout, short-creeping. Scaly.

Fronds: 70 to 100 cm long. Clustered. Leathery. Lustrous golden green.

Stipe: 25 to 40 cm long. Lower part with large, tan scales.

Rachis: Tan, slightly scaly.

Blade: 45 to 60 cm long and 20 to 35 cm wide. Oblong-ovate, with short, pointed tip. Bipinnate to pinnate-pinnatifid.

Pinnae: Lowest pair opposite, others alternate. Deeply cut. Oblong with long, pointed tips. Pinnules oblong with pointed tips and variously incised margins. Fertile pinnae on upper part of blade.

Sori: Medial. In rows on each side of pinnule midvein. Indusia peltate, kidney-shaped.

Habitat: Rich, rocky woods and wooded slopes.

Range: Rare, but often found where both parents are present. Eastern North America southward to North Carolina and northern Georgia and westward to Arkansas and Illinois. In Georgia, only in Union County.

Remarks: This is a sterile hybrid of *D. goldiana* and *D. marginalis.*

Crested Wood Fern

Dryopteris cristata (L.) A. Gray

Name: In 1753 Linnaeus named this fern *Polypodium cristatum,* never explaining the significance of the epithet, which means "crested." Asa Gray transferred it to *Dryopteris* in 1848.

Rootstock: Stout, ascending. Covered with light brown scales.

Fronds: Fertile, 40 to 75 cm long, erect, deciduous. Sterile, 20 to 60 cm long, spreading, evergreen. Firm-textured, bluish green.

Stipe: Fertile, 10 to 25 cm long. Sterile, 5 to 15 cm long. Green above, dark at base. Sparse, light brown scales.

Rachis: Green. Slightly scaly on lower part.

Blade: Fertile, 30 to 50 cm long and 8 to 15 cm wide. Sterile, 15 to 45 cm long and 6 to 12 cm wide. Narrowly oblong. Pinnate-pinnatifid.

Pinnae: Most opposite, short-stalked. Basal widely spaced, triangular. Apical closer and more oblong. All broadest near stipe; short, tapering to blunt tip. Segments with serrate margins. Pinnae of fertile fronds twisted at right angles to plane of blade (like an open Venetian blind).

Sori: Round, medial. Indusia kidney-shaped, glabrous.

Habitat: Wet woods, wooded swamps, marshes, and bogs.

Range: Europe. Northern and eastern North America southward to North Carolina and Tennessee. Found in southern Fulton County, Georgia, in 1984.

Florida Wood Fern

Dryopteris ludoviciana (Kunze) Small

Name: Buckley collected this fern in Florida in 1842 and Hooker described it in 1850 as *Nephrodium floridanum.* Otto Kuntze transferred it to *Dryopteris* in 1891. In 1938 Small pointed out that its earliest name is *Aspidium ludovicianum,* given by Gustav Kunze in 1848, and made the proper transfer. *Ludoviciana* means "of Louisiana," where it was first discovered in the early 1800s. It is also commonly called "southern shield fern" or "southern wood fern."

Rootstock: Stout, creeping, with tan scales.

Fronds: 60 to 120 cm long. Produced in rows. Dark green, lustrous and leathery. Evergreen.

Stipe: 15 to 30 cm long with tan scales.

Rachis: Grooved, with tan scales.

Blade: 45 to 90 cm long and 15 to 30 cm wide. Fertile longer than sterile. Elliptic-lanceolate. Fertile broadest near middle, where sterile and fertile pinnae meet. Pinnate-pinnatifid.

Pinnae: Long-oblong with pointed tips. Deeply incised. Sori on pinnae of upper half of fertile blade only. Fertile pinnae much narrower than sterile and noticeably contracted. Prominent veins run to edges of toothed pinnules.

Sori: Round, medial. Indusia peltate, kidney-shaped.

Habitat: Swamps, damp woods, shaded limestone outcrops, and margins of cypress swamps.

Range: North Carolina southward and westward along the Coastal Plain to central Florida, Alabama, and Louisiana. In Georgia, confined to the Coastal Plains and nowhere common.

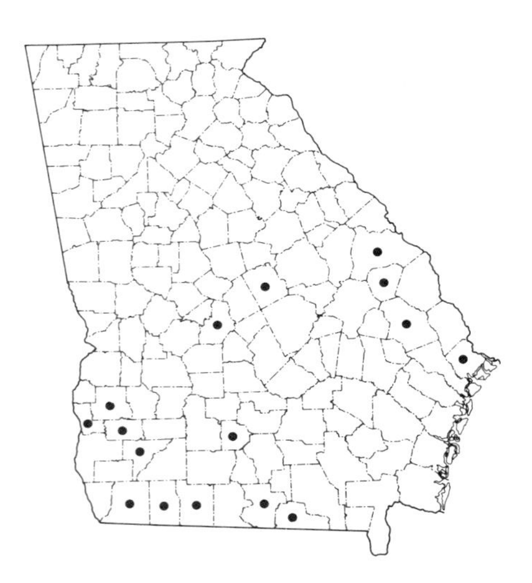

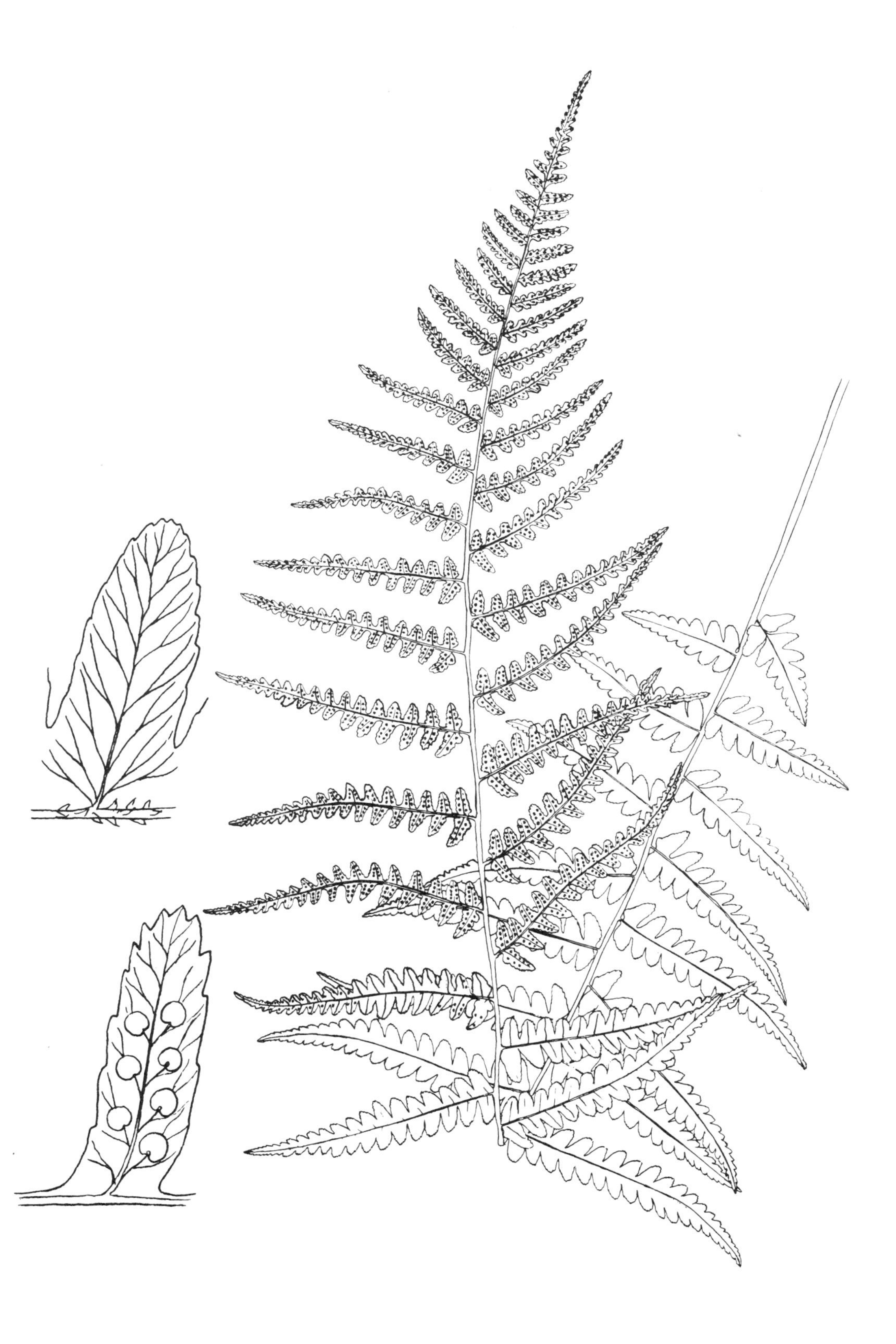

Hybrid Wood Fern

Dryopteris × *australis* (Wherry) Small

Name: In 1937 Wherry named this taxon *Dryopteris clintoniana* var. *australis,* which Small in 1938 recognized as a hybrid. The epithet means "southern."

Rootstock: Stout, creeping, with tan scales.

Fronds: 90 to 130 cm long. Dark green, leathery. Evergreen.

Stipe: 30 to 40 cm long. Tan, covered with tan-to-brown scales.

Rachis: Tan. Lower portion very scaly, upper less so.

Blade: 60 to 90 cm long and 20 to 30 cm wide. Three times longer than wide. Widest in middle. Narrowly elliptical with short, tapering tip. Pinnate-pinnatifid.

Pinnae: Alternate. Dimorphic. Upper fertile pinnae somewhat narrower and more widely spaced than lower sterile ones. Fertile pinnae approximately one-half of blade. Pinnules oblong with rounded tips.

Sori: Medial, in rows on each side of pinnule midveins. Indusia peltate, kidney-shaped.

Habitat: Swampy woods.

Range: Very rare. Found only in North and South Carolina, Alabama, Georgia, and Louisiana, although sometimes locally in large colonies.

Remarks: This is a sterile hybrid of *D. celsa* and *D. ludoviciana.*

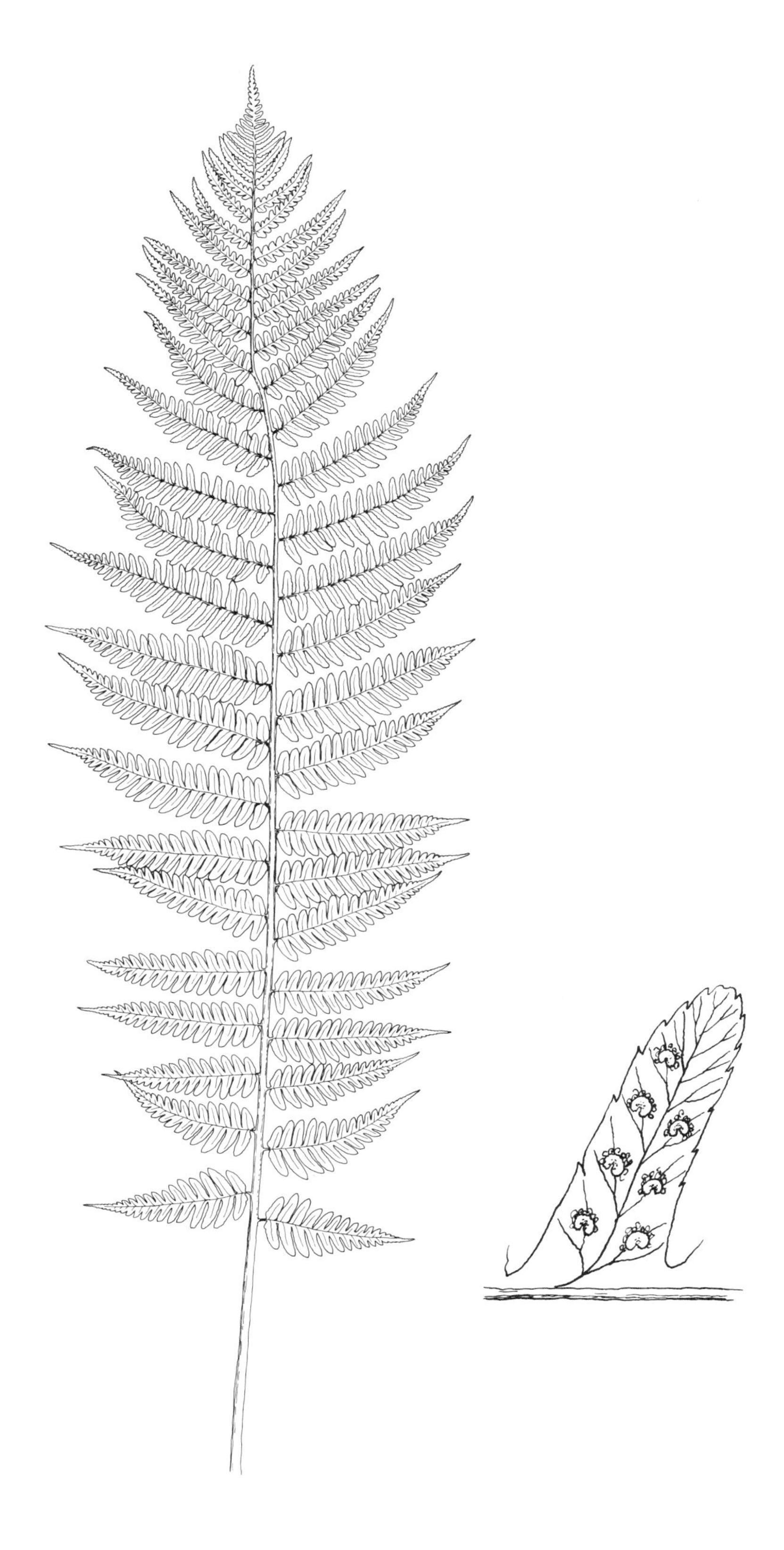

Goldie's Wood Fern

Dryopteris goldiana (Hook.) A. Gray

Name: John Goldie, a British traveler of the early nineteenth century, discovered this taxon near Montreal. Sir William J. Hooker, an eminent British botanist, named it *Aspidium goldianum* in his honor in 1822. Asa Gray transferred it to *Dryopteris* in 1848.

Rootstock: Stout, woody, short-creeping. Dark scales.

Fronds: 90 to 120 cm long. More or less scattered. Broad, lustrous golden green.

Stipe: 30 to 40 cm long. Tan. Basal scales many, large, very dark and shiny or tan with dark brown centers.

Rachis: Tan. Scales small, light brown.

Blade: 60 to 80 cm long and 20 to 40 cm wide. Oblong-triangular with tip narrowing abruptly. Pinnate-pinnatifid. Leathery, backward tilting.

Pinnae: Alternate. Oblong with pointed tips. Deeply cut. Pinnule margins with low, appressed teeth.

Sori: Round. In two rows close to midvein. Indusia peltate, kidney-shaped.

Habitat: Cool, rich woodlands, including rocky limestone woodlands.

Range: Maritime provinces to the Great Lakes states and southward along the Appalachians to northern Alabama and northernmost Georgia.

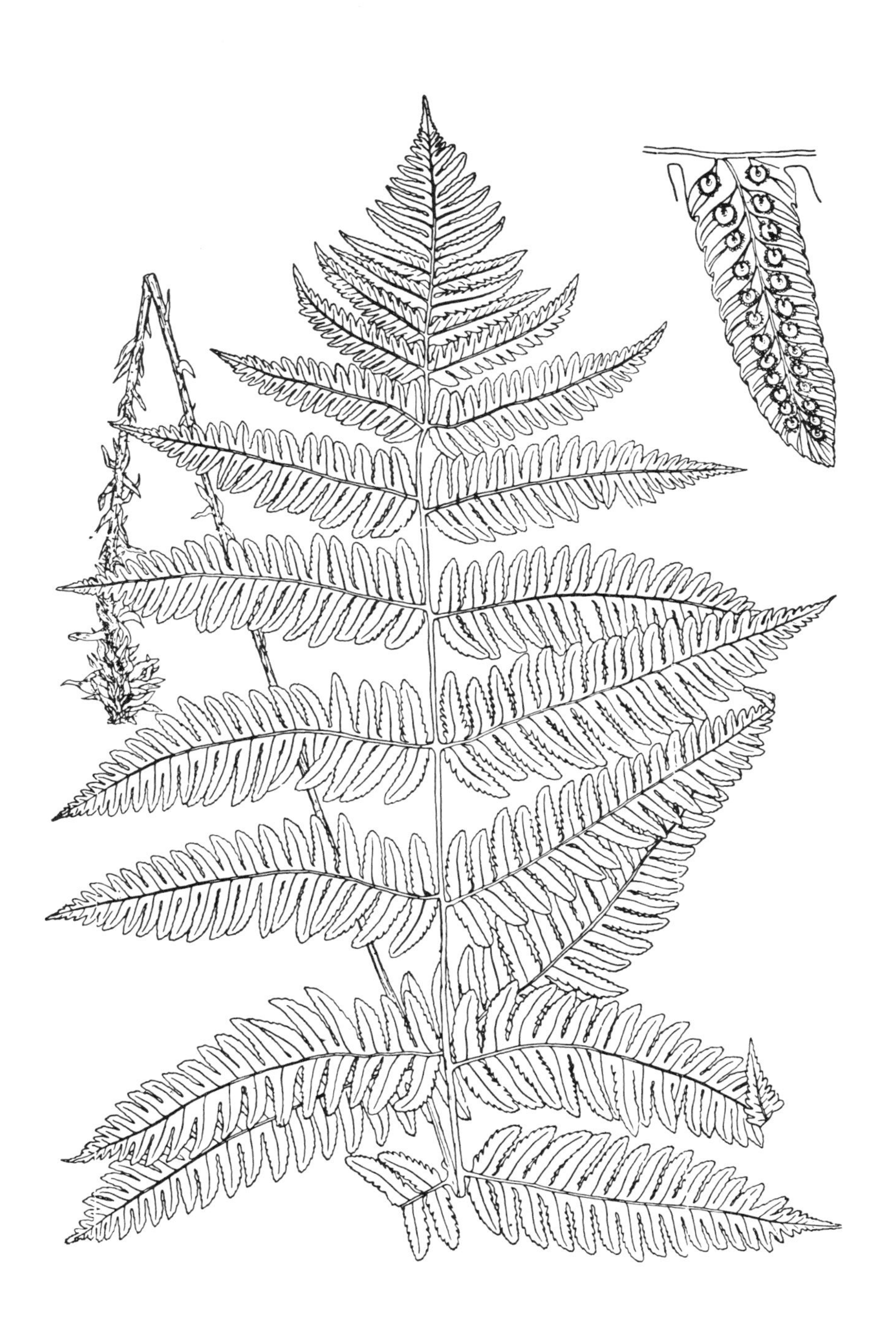

Log Fern

Dryopteris celsa (W. Palmer) Small

Name: The log fern was first discovered in the Great Dismal Swamp in Virginia in 1899 and was classified that year by William Palmer as a subspecies of *Dryopteris goldiana.* In 1938 Small made it a separate species. The Latin name *celsa* means "held high" and refers to the common growth habit above ground on logs or humus. The common name, "log fern," likewise refers to this habit.

Rootstock: Shallow, short-creeping. Covered with dark scales.

Fronds: 90 to 120 cm long. Firm in texture. Evergreen.

Stipe: 30 to 40 cm long. Light green. Covered with a mixture of broad and narrow, pale brown scales.

Rachis: Green. With smaller and fewer scales than on stipe.

Blade: 60 to 80 cm long and 20 to 30 cm wide. Oblong, slightly narrowed at base and gradually tapering at tip. Pinnate-pinnatifid.

Pinnae: Lanceolate; long, tapering, and deeply lobed. Lower pinnae narrowed at base.

Sori: Round. Medial to near midveins. Indusia peltate, kidney-shaped. Spores mature in summer and fall.

Habitat: Rotted logs and humus piles near large trees in hammocks and near swamps. Soil moderately to strongly acidic.

Range: Rare in widely scattered sites in the eastern United States from New York and Michigan southward to Louisiana. Mostly in the Great Dismal Swamp of Virginia and North Carolina. In Georgia, only in Dade and Walker counties.

Remarks: In North America the genus *Dryopteris* is composed of 13 fertile species and 29 sterile hybrids. Several of the fertile species are themselves hybrids in origin. Among these latter is *Dryopteris celsa,* whose parents are *D. ludoviciana* and *D. goldiana.*

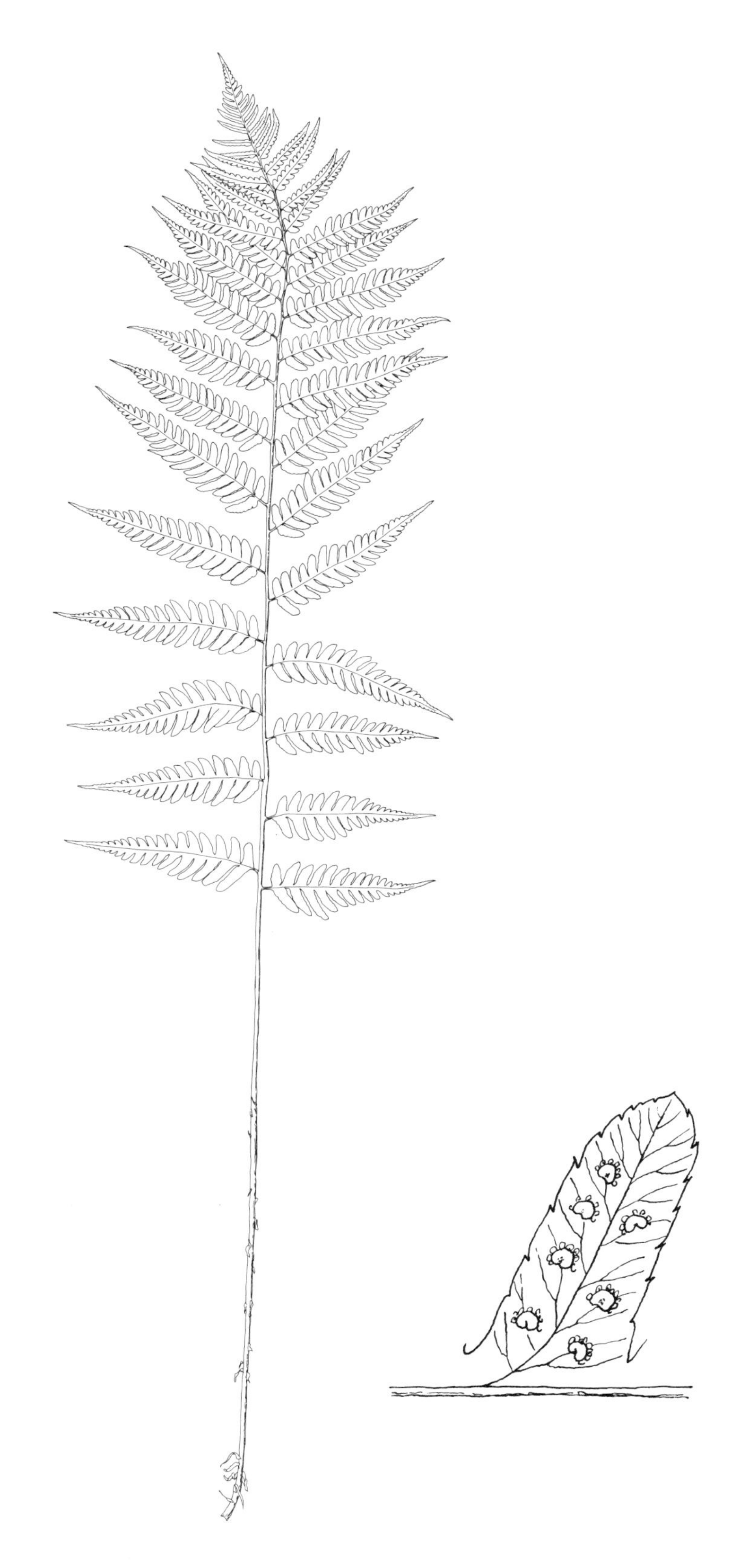

Evergreen Wood Fern

Dryopteris intermedia (Muhl. ex Willd.) A. Gray

Name: In 1810 Willdenow published the name *Polypodium intermedium* from Muhlenberg's description of a Pennsylvania plant. Gray transferred it to *Dryopteris* in 1848. This fern is also commonly known as "spiny wood fern," "glandular wood fern," and "fancy wood fern."

Rootstock: Thick, semierect, covered by persistent leaf bases. Scales light brown with dark centers.

Fronds: 45 to 90 cm long. Clustered. Lacy, glossy, dark green. Evergreen, but become ragged before spring.

Stipe: 15 to 30 cm long. Covered with light brown scales.

Rachis: Straw-colored. Scaly. Upper with pinhead glands.

Blade: 30 to 60 cm long and 15 to 25 cm wide. Oval to narrowly triangular with parallel sides. Tapers gradually to tip. Bipinnate-pinnatifid to tripinnate.

Pinnae: Lanceolate. Pinnules cut into sharp, spreading teeth. Basal pinnules on lower side of lowest pinnae no longer or wider than opposite, upper pinnules and shorter than next, adjoining pinnules.

Sori: Small, round, medial. Indusia peltate, kidney-shaped with minute pinhead glands.

Habitat: Various rich, woodland soils, including swampy and rocky places.

Range: Widespread in eastern temperate North America, southward to northern Alabama and Georgia. Comparatively rare in Georgia—in only some of the northernmost counties.

Remarks: *D. intermedia* is sometimes confused with the closely related species *D. carthusiana*, which has not been found in Georgia. [Note: *D. carthusiana* (Vill.) H. P. Fuchs was known as *D. spinulosa* until 1980, when Fraser-Jenkins concluded that Villars's older name applied.] *D. carthusiana* is not evergreen and not glandular. On it, the first downward-pointing pinnules of the basal pinnae are often 2 times longer than their opposite upward-pointing pinnules, and the downward-pointing pinnules next to the rachis are always the longest.

Christmas Fern

Polystichum acrostichoides (Michx.) Schott

Name: In 1803 André Michaux first named this plant *Nephrodium acrostichoides,* the epithet meaning "like *Acrostichum,*" another genus of ferns. In 1834 Schott transferred it to *Polystichum,* a genus established by Roth in 1799. The name means "many rows," referring to the sori. The common name, "Christmas fern," refers to its having been used by early New England settlers as Christmas decorations.

Rootstock: Ascending, stout, scaly.

Fronds: 30 to 70 cm long. Clustered. Evergreen through winter. Sterile shorter than fertile. Lustrous green.

Stipe: 10 to 20 cm long. Round, brown at base, green above. Stout with large, tan scales.

Rachis: Green, scaly.

Blade: 20 to 50 cm long and 7 to 13 cm wide. Lanceolate, widest above base. Pinnate.

Pinnae: Oblong-lanceolate. About 1 cm wide. Short-stalked, upper auricle. Margins bristle-toothed. Fertile pinnae only on upper one-third to one-half of fertile frond. Fertile pinnae markedly smaller and narrower than sterile.

Sori: Round, usually in rows on each side of pinna midrib. Distinct at first, becoming confluent and covering entire back of pinnae. Indusia peltate.

Habitat: Moist woods, shaded slopes and ravines. Usually in partial shade, but in all sorts of soils.

Range: Abundant throughout eastern North America. One of the most common ferns of northern Georgia, and frequent throughout southern Georgia except in flat, sandy regions of the southeastern part of the state.

Remarks: Occasionally a distinctive variant of this fern will be found—known as f. *incisum* (Gray) Gilbert. In f. *incisum* the pinnae are more deeply and conspicuously toothed. The pinnae may be enlarged and somewhat triangular in outline, and a number of the "sterile" pinnae may have fertile tips. Wagner, Farrar, and McAlpin in their study of the pteridology of the Highlands area in North Carolina determined that fronds of this form appear after the normal spring flush, will be found on plants with normal leaves, and have no taxonomic significance.

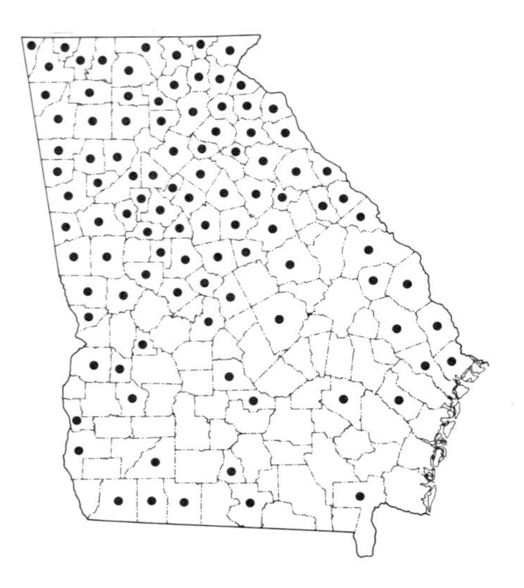

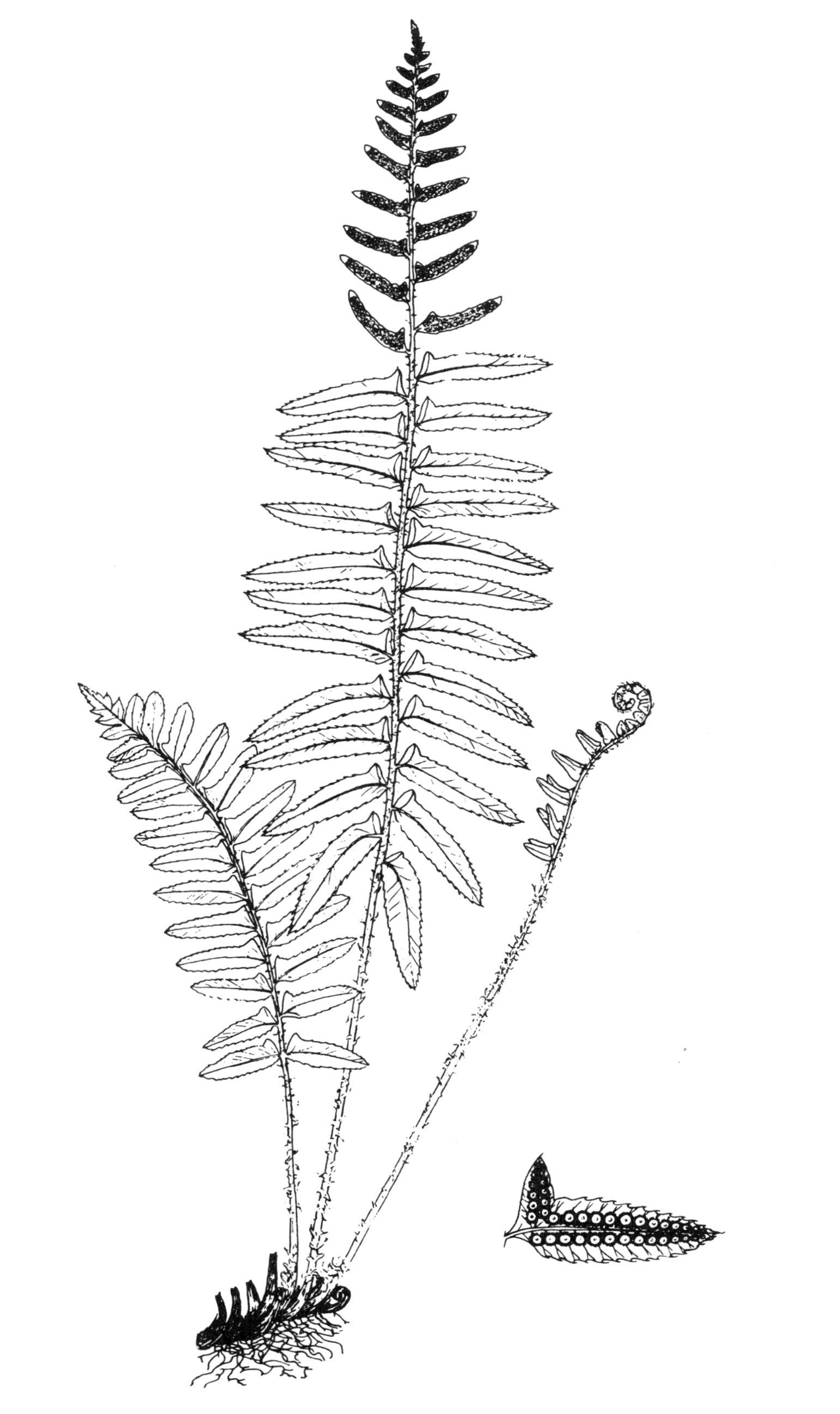

Japanese Holly Fern

Cyrtomium falcatum (L. f.) Presl

Name: Linnaeus's son first named this fern *Polypodium falcatum* in 1781. In 1836 Presl established the new genus *Cyrtomium,* including this taxon. *Falcatum* means "scythe-shaped" and *Cyrtomium,* "cut in a curve," both terms being aptly descriptive of the pinnae of this fern. It is also commonly called "Asian holly fern."

Rootstock: Short, stout, covered with conspicuous, dark brown scales.

Fronds: 28 to 55 cm long. Relatively few, forming a wide vaselike crown. Dark green above, paler beneath. Leathery texture.

Stipe: 10 to 20 cm long. Stout. Densely covered with brown scales.

Rachis: Smooth. Green with few hairs.

Blade: 18 to 35 cm long and 8 to 15 cm wide. Pinnate.

Pinnae: 4 to 10 pairs. Large, close-set, and short-stalked. Obliquely ovate with long, curving, acuminate tip. 2 to 3 cm wide with coarsely toothed margins like a holly. Veins form complex networks and are particularly conspicuous on lower side.

Sori: Large, round, scattered, lying along or at ends of veins with areoles. Indusia prominent, peltate, persistent.

Habitat: Circumneutral soil in clayey banks, rubble, and masonry walls.

Range: Commonly cultivated as a house or garden plant in mild climates. Escaped from cultivation in scattered localities in the southeastern United States. In Georgia, in Glynn, Colquitt, and Chatham counties.

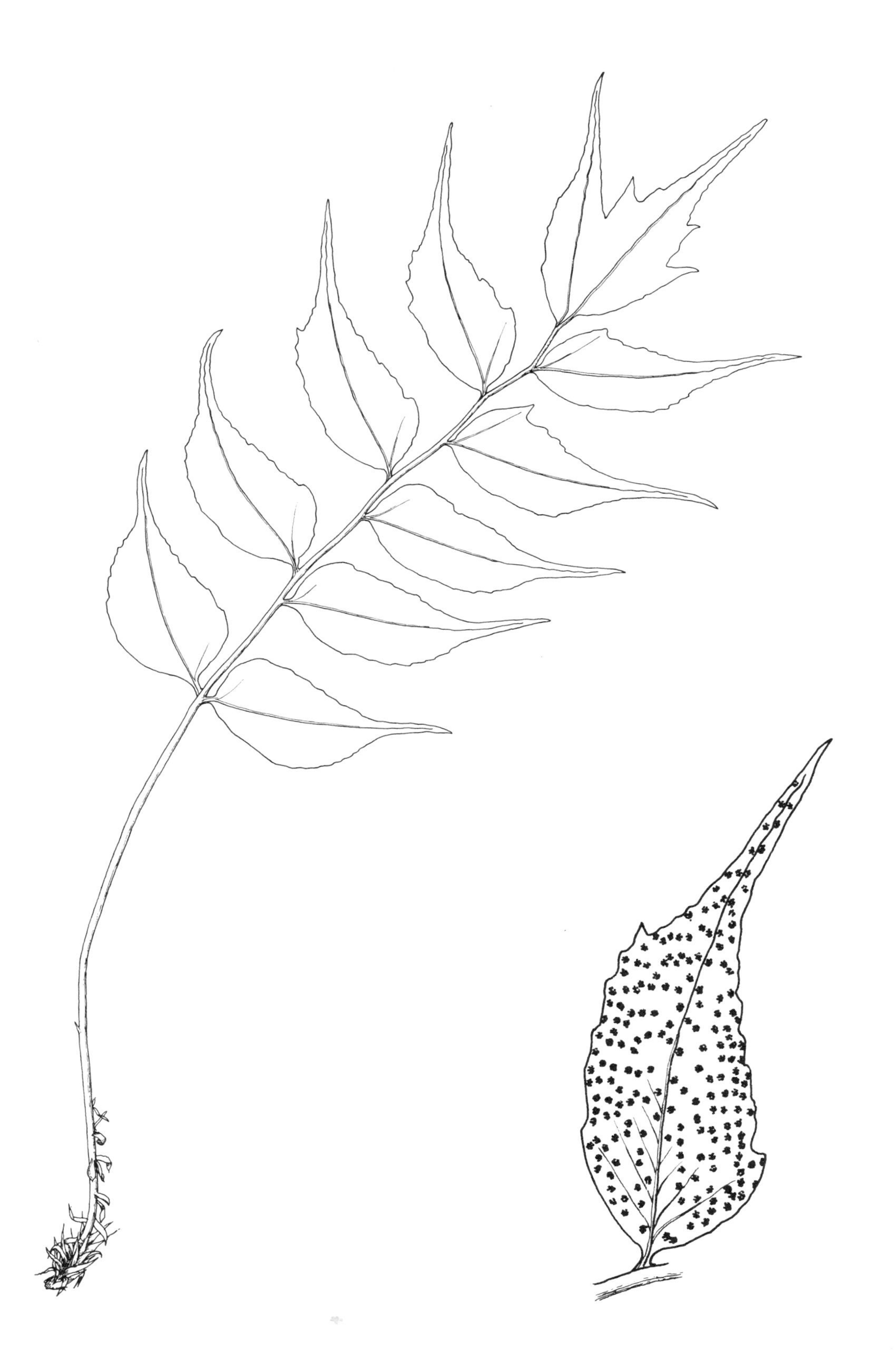

Holly Fern

Cyrtomium fortunei J. Smith

Name: John Smith, a British horticulturist of the nineteenth century, named this taxon in 1866 in honor of Robert Fortune, who had brought a specimen to England from China, where it is common.

Rootstock: Stout, ascending. Covered with dark, lanceolate to ovate, acuminate scales.

Fronds: 40 to 75 cm long. Arise in wide, vaselike cluster. Evergreen.

Stipe: 8 to 15 cm long. Green, densely covered with large, light and dark brown scales.

Rachis: Green, grooved, slightly scaly above. Brown and very scaly below.

Blade: 32 to 60 cm long and 12 to 18 cm wide. Pinnate. Dull green above, light green below. Leathery texture.

Pinnae: Alternate. Lanceolate, sickle-shaped, 2 cm broad at base. 9 to 20 pairs, distinctly stalked, slightly ascending. Long, triangular, terminal pinna. Margins serrate. Prominent midveins shiny black and grooved above, light green and raised below. Veins reticulate.

Sori: Small, round, distant, and scattered over all or part of lower surface of fertile pinnae. Indusia peltate.

Habitat: Circumneutral soil, invading masonry, rubble heaps, and clayey banks.

Range: Escaped from cultivation and naturalized in sites from South Carolina to Florida. Found in three Georgia counties.

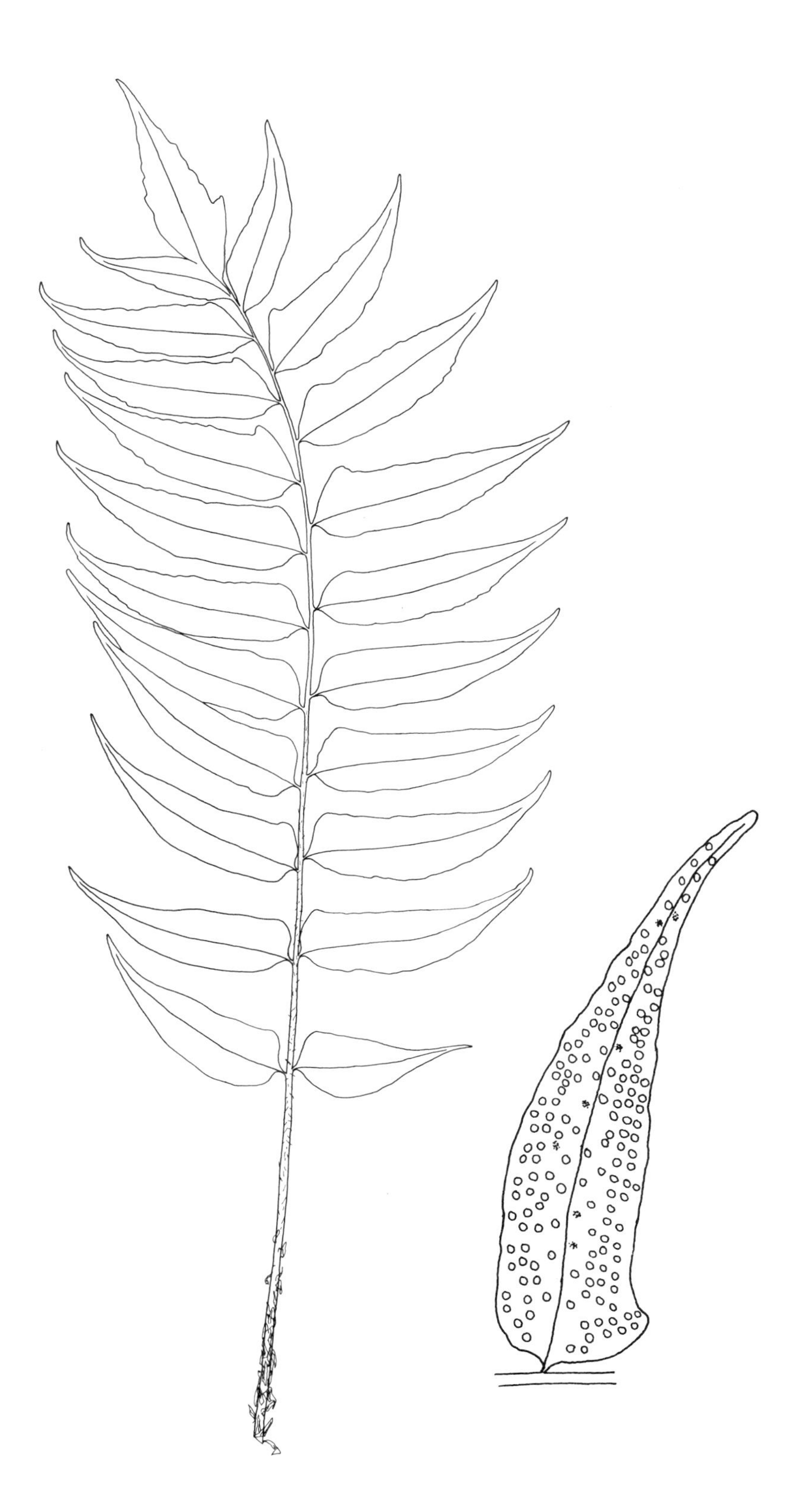

American Pillwort

Pilularia americana A. Br.

Name: In 1753 Linnaeus named this genus from the Latin word for "little ball," referring to the small, globular sporocarps. Alexander Braun recognized the distinctiveness of the American species and named it in 1864.

Rootstock: Naked, slender, wide-creeping, forming colonies.

Fronds: 2 to 10 cm long. Small, resembling grass or rush. No leaf blades or leaflets. Young uncurl at tips like other ferns.

Sporangia: Borne in globular, nutlike sporocarps at base of leaf stalks. Brownish, covered with short, appressed hairs. Sporocarps about 0.25 cm in diameter.

Habitat: Muddy and seasonal pools and wet soil by ponds.

Range: Peculiarly distributed in a few central and western states. Found east of the Mississippi River only in Georgia.

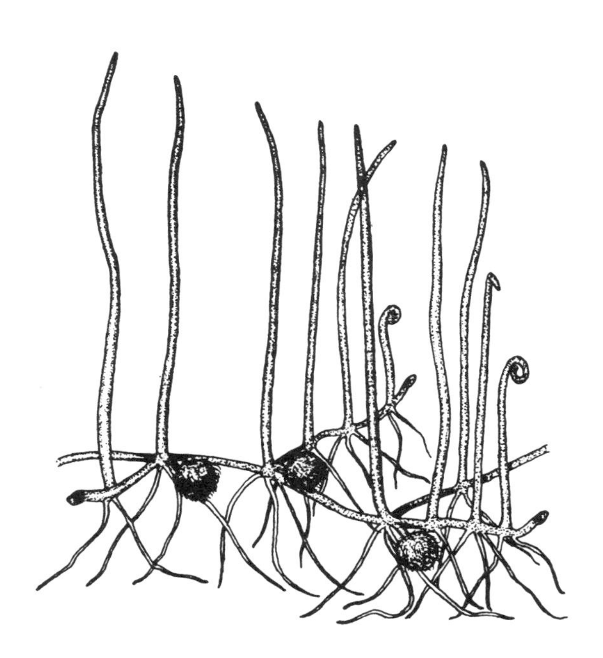

Mosquito Fern

Azolla caroliniana Willd.

Name: J. B. Lamarck, a famous French naturalist, established the *Azolla* genus in 1783. *Azolla* comes from the Greek meaning "killed by drought," referring to the fern's aquatic habitat. Willdenow gave the species name to this fern in 1810 based on material from "Carolina." It is commonly called "mosquito fern" because a dense growth of it covers water so thickly as to exclude mosquito larvae.

Roots: Up to 5 cm long. Threadlike, numerous, lax, hanging from axils of stems.

Stems: Hairlike, free-floating on water surfaces. Branched by equal forking, forming small, spreading, roundish to ovate-oblong plants about 1 cm in diameter. Though individual plants are the tiniest of all ferns or fern allies, mats of *Azolla* may completely cover sections of water.

Leaves: Tiny (about 0.8 mm long and 0.6 mm wide). Overlapping, in two rows along stems. Each composed of two round lobes. Upper lobe emersed, oval, deep green-blue to green to purplish red. Lower lobe submerged, inconspicuous, thin, and colorless.

Sporangia: Plants heterosporous. Male and female sporangia produced in separate sporocarps borne in axils of leaves. Glochidia one-celled.

Habitat: Floating in still water of ponds, pools, ditches, and backwaters. Also on mud along margins of pools where water has receded.

Range: Frequent in the eastern United States and the West Indies. In the Southeast, mostly along the Atlantic and Gulf coasts. In Georgia, in scattered sites in the Coastal Plains.

Remarks: The minute male spores of *Azolla* species join to form jellylike masses with protruding arrowlike hairs, called glochidia. Species identification is based on details of the glochidia and requires a microscope for examination. As fertile plants are rarely found, species identification is usually difficult.

Azolla may possibly be not only the most important fern in the world economically but among the major economic plants because of its relationship to the culture of rice. All species of *Azolla,* so far as we know, contain within the spaces between their leaf lobes a distinctive blue-green alga, *Anabaena azollae,* which is capable of fixing nitrogen and thus provides nutrients for the growth of rice in paddies.*

* T. A. Lumpkin and D. L. Plucknett, *Azolla as a Green Manure: Use and Management in Crop Production* (Boulder, Colorado: Westview Press, 1982).

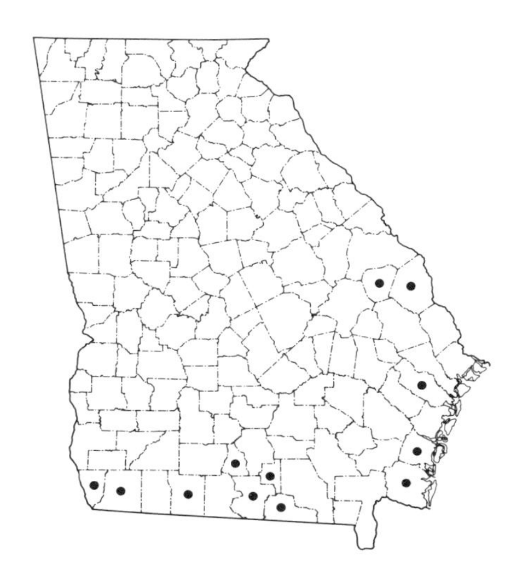

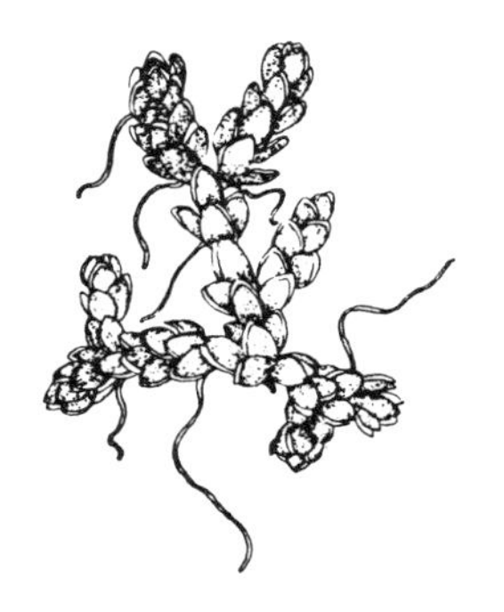

Mosquito Fern

Azolla filiculoides Lam.

Name: Lamarck named this species in 1783. The epithet means "like a little fern."

Roots: Up to 5 cm long. Threadlike, numerous, lax.

Stems: Hairlike, free-floating in water. Branching laterally to form small, roundish plants 2 to 5 cm long. Though individual plants are small, mats may completely cover sections or bodies of water.

Leaves: Tiny, with minute, short hairs. Overlapping in two rows along stems. Each composed of two round lobes. Upper lobe emersed, oval, deep green-blue to green and purplish red. Lower lobe submerged, inconspicuous, thin, and colorless.

Sporangia: Plants heterosporous. Male and female sporangia produced in separate sporocarps borne in axils of leaves. Glochidia mostly one-celled, but rarely with one or two cross-walls near the tip.

Habitat: Floating on still water of quiet streams and ponds.

Range: Western North America, Central and South America. Occasionally escapes elsewhere, as in McIntosh County, Georgia.

Remarks: See *Azolla caroliniana.*

side view of double leaves

Water Spangles

Salvinia minima Baker

Name: Michel Adanson established this genus in 1763, naming it in honor of the seventeenth-century Italian naturalist A. M. Salvini. John Baker in 1886 named this species *minima,* meaning "smallest." It is also commonly called "water fern," "floating fern," and "South American pond fern."

Stems: 4 to 6 cm long, little-branched, light brown. Brittle, breaking to form new plants. No roots.

Leaves: About 1 cm in diameter with short petiole. Opposite in two rows along stem. Entire, round or oval with cordate (heart-shaped) base. Blue-green. Upper surfaces covered by tiny hairs with four spreading branches.

Sporocarps: Plants are heterosporous. Sporangia are contained in sporocarps—hard, nutlike structures. Male and female sporocarps are same size and shape. Borne on trailing, dissected, rootlike organs that are actually leaves.

Habitat: Floating on water surface or creeping on mud. On waters of quiet ponds and backwaters and slow-moving streams. Also on mud in marshes and along edges of ponds where water has receded.

Range: Tropical America and the southeastern United States.

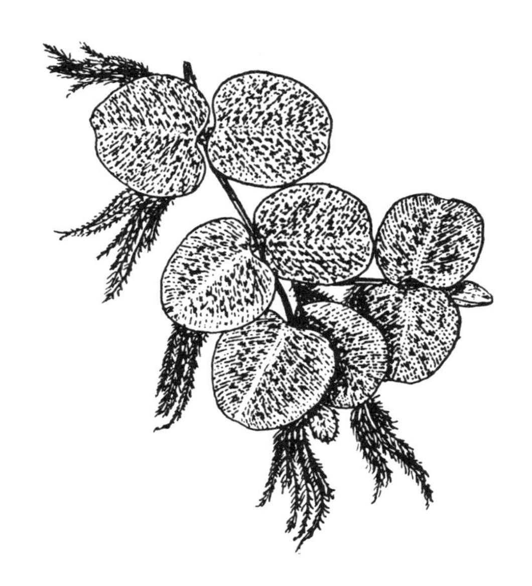

Other Pteridophytes

The other pteridophytes, or so-called fern allies, are not closely related to the ferns. Like ferns, however, they are vascular plants that are reproduced by spores. Unlike true ferns, they do not have large leaves, or megaphylls, with branching vein systems. Instead, their smaller, scalelike leaves, or microphylls, have only a single, unbranched vein. These other pteridophytes include the horsetails (*Equisetum*), quillworts (*Isoetes*), clubmosses (*Lycopodium*), whisk fern (*Psilotum*), and spikemosses (*Selaginella*).

Ancestors of the present-day ferns and other pteridophytes were among the earliest of this earth's plants. Some 300 million years ago, horsetails, clubmosses, tree ferns, and cone-bearing trees grew together in vast swamps. Their remains formed mats of plant tissues that were covered and compressed by sediments, were chemically changed, and became the huge deposits of coal we have today.

The genera of non-fern pteridophytes are so different from ferns and from each other that they will be described below briefly. A number of these plants are so peculiar that most field naturalists (even botanists) have never seen them.

LYCOPODIUM

Linnaeus established *Lycopodium* as a genus in 1753, using an ancient Greek name meaning "wolf's foot." Although called "clubmosses," they are not related to nor resemble mosses. The plants are small, perennial, usually evergreen, of upright, trailing or creeping growth. They often look like small pines, hemlocks, or cedars, which they are commonly called, and are often found in coniferous woodlands.

The rootstocks, or horizontal stems, grow along the ground, either above or below it, and have rather coarse roots. These elongate by apical growth, often forming long runners from which a number of separate, upright branches arise. This vegetative method of reproduction frequently produces large colonies in meadows and woodlands. Some species also reproduce by bulblets, or plantlets, called *gemmae.* These bulblets appear at the bases of the upper leaves and when fully developed fall to the ground and start new plants.

Arising from the rootstocks are the erect or arching stems, covered with multiple rows of small, scalelike leaves. These leaves are arranged spirally around the stem, either appressed or spreading. They are narrow and pointed, with a single vein.

The spores of clubmosses are in globular or elliptical sporangia, which are borne near the base of specialized leaves called *sporophylls.* These may be in zones along the stems or branches, or grouped in terminal cylindrical cones. These cones, or *strobili,* may be at the tip of a regular, erect branch or on a special stem, the *peduncle.*

The spores are yellow, and have been used for coating pills and in soothing powders for chafes and wounds. As they give off a flash explosion when ignited, they have also been used in fireworks and photographic flashes.

SELAGINELLA

The genus *Selaginella* was established in 1805 by the French pteridologist Ambroise Palisot de Beauvois. The name means "small *selago,*" an ancient name for a small firlike plant. The common name, "spikemoss," comes from the spikelike cones of the sporophylls.

The plants are small, creeping, and mosslike, growing erect or flat in tufts or mats. The leaves are very small (usually about 2 to 5 mm long), round to linear, with single, unbranched midveins, which often end in a hairlike tip, or *seta.*

The sporangia are borne in the axils of the sporophylls, which are clustered in a sessile strobilus, or cone. This strobilus is usually four-sided and located at the end of the fertile branch.

Selaginellas have both microspores and megaspores. The former are usually numerous; the latter, 1 to 4 per megasporangium. The megaspores are irregularly spherical with a flattened base and three triangular faces separated by ridges. Megaspores are 230 to 450 microns in diameter. The surfaces are weakly or deeply grooved in a network pattern, pebbled, or smooth. The microscopic examination of the irregularities in megaspore surface ornamentation is often of considerable help in the identification of *Selaginella* species.

ISOETES

The genus of quillworts, or Merlin's-grass, was named *Isoetes,* meaning "equal points," by Linnaeus in 1753. The common name of "quillwort" refers to the supposed resemblance between their leaves and the quill of a feather.

The plants are small, evergreen, generally solitary, resembling grass or onion-like tufts. Though sometimes terrestrial, they are mostly amphibious and often totally submerged.

The stem is cormlike, submerged, usually with two lobes. The fibrous roots grow down from the corm, originating from a groove between the lobes. They are numerous, generally forked (dichotomous), and monomorphic.

The sporophylls, or spore-bearing leaves, vary in number from 5 to 100 or more in individual plants and from 2 to 60 cm in length. They are borne on top of the stem, are long, slender, and grasslike. The leaves, however, are not flat in cross section like grass but are crescent-shaped, convex on the outer side and concave on the inner. The arrangement of the leaves is usually spiral around the corm. They grow imbricated, or overlapping, with the youngest leaves on the inside. They have a single vein, but occasionally have veinlike thickenings called *peripheral strands* parallel to the vein. Depending on the species, they may have *stomata,* or breathing pores. The leaves have four vertical air channels that are irregularly interrupted by horizontal walls, giving a jointed appearance. At the base of the leaf is a tiny flap, termed the *ligule.*

Sporangia are located in the top side of swollen leaf bases and are wholly or partially covered by a *velum,* a membrane formed by thin leaf tissue. The sporangia may or may not have pigmented cells in the walls.

Isoetes is heterosporous—having both male microspores and female megaspores. The former are usually numerous; the latter, moderate in number. Both spore types are borne in different sporangia on the same plant. The megaspores are relatively large—300 to 800 microns in diameter, or about the size of a printed period. Their surfaces contain one convex and three triangular faces separated by ridges. The areas between the ridges are of 4 types: reticulate, or pitted or netted; tuberculate, or pebbled; echinate, or with prickly spines; and cristate, or with crested appendages.

Although some identification of species is possible from vegetative forms, the classification of *Isoetes* is largely based on megaspore ornamentation that requires magnification of some 30 to 50 times. Currently, less is known about the taxonomy of *Isoetes* than of any of the other pteridophytes.

EQUISETUM

Linnaeus established this genus in 1753 using an ancient plant name that literally means "horse bristle." This name and the common "horsetail" are appropriate as some of the plants resemble a horse's tail. The other common name, "scouring rush," was adopted because silica deposits in the plants make them useful for scouring purposes.

The plants have an extensive perennial underground system. The upright stems are cylindrical, bamboolike with longitudinal grooves and ridges often roughened by silica deposits. The stems may be branched or unbranched, all alike or dimorphic. At conspicuous joints or nodes numerous small, narrow, sharp-pointed, scalelike leaves are whorled and fused to form the *leaf-sheath.*

The aerial stem and branches are hollow with a large central canal, or *centrum,* surrounded by smaller *vallecular canals* under the grooves and *carinal canals* under the ridges. The sizes and shapes of these three kinds of cavities in cross section provide keys to determining the species.

Spores are produced in cones, or strobili, at the tip of each fertile stem. These cones are composed of many umbrella-like structures, the *sporangiophores,* under which the sporangia are attached. Spores are green, minute, numerous, and round or oval. Attached at one spot to each spore are four ribbon like strips with spatulate tips called *elaters.* Reacting to changes in humidity, these curl and uncurl around the spore to aid in spore dissemination.

Shining Clubmoss

Lycopodium lucidulum Michx.

Name: André Michaux, traveling in North America, found and named this taxon in 1803. *Lucidulum,* meaning "somewhat shiny," refers to the slightly lustrous quality of the leaves.

Horizontal Stem: Short-creeping and rooting. Obliquely rising to form erect stems.

Erect Stems: 15 to 25 cm high. Evergreen. Growing in clumps. Short sporophylls and longer sterile leaves in alternating bands, giving a knobby effect.

Sterile Leaves: 0.5 to 1.5 cm long. Shiny dark green. In 6 rows. Lance-shaped. Broadened and serrate above middle.

Sporophylls: Shorter, narrower, and less serrate than sterile leaves. Sporangia borne in axils. Gemmae, asexual reproductive bodies with broad lobes, are formed among upper leaves. These may form new plants vegetatively.

Habitat: Moist woods, wet rocky slopes, mountain ravines, stream banks, and swamp margins in acidic or circumneutral soil rich in humus.

Range: Common throughout most of the upland forest of eastern North America from eastern central Canada westward to Minnesota and Arkansas and southward to northern Alabama and Georgia. Found in Georgia in the higher mountains of a number of northeastern counties.

Remarks: This species hybridizes with *L. selago* and *L. porophilum.*

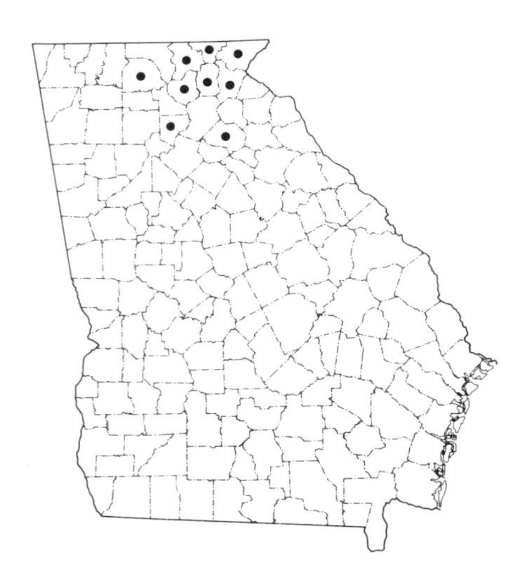

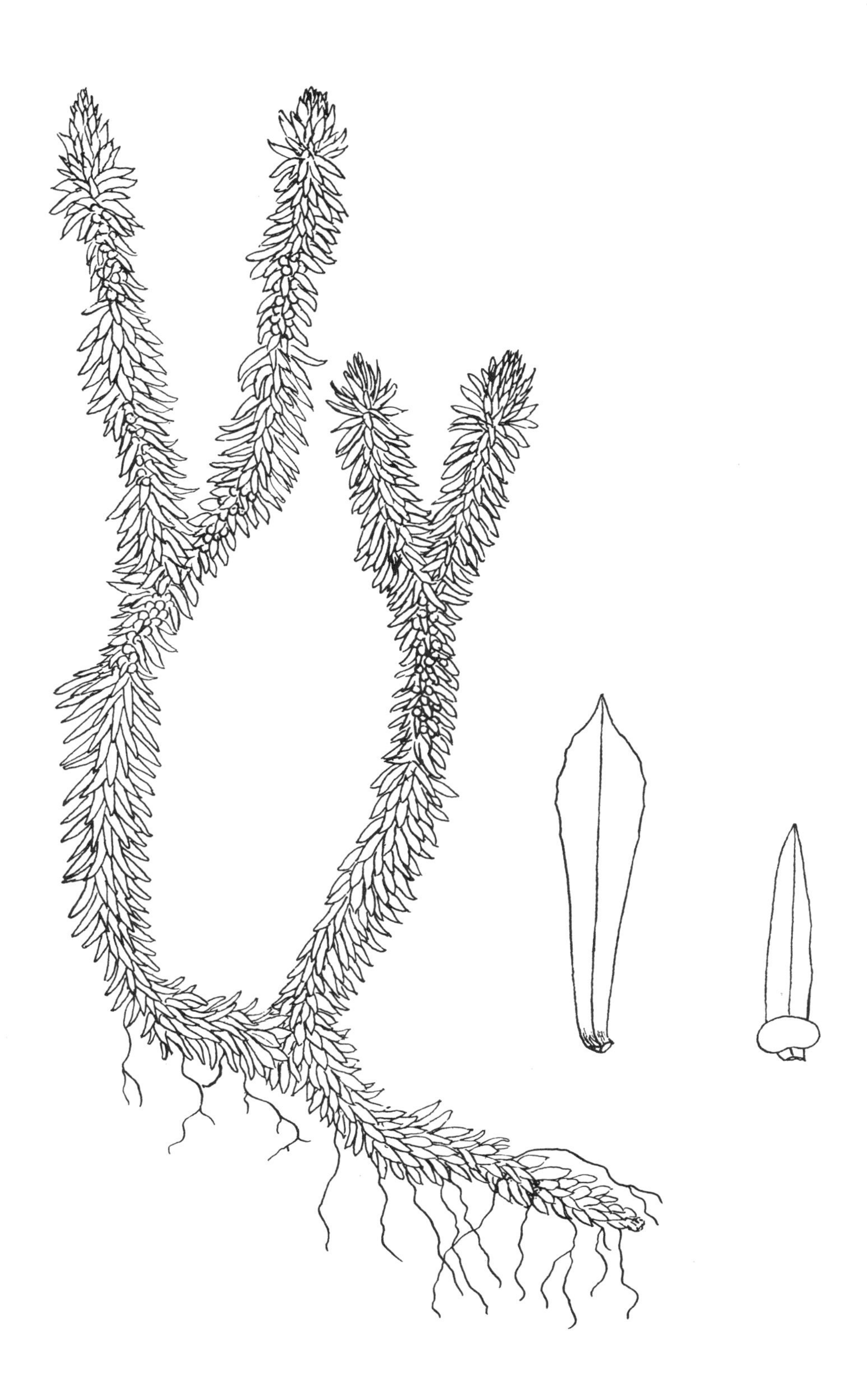

Fir Clubmoss

Lycopodium selago L.

Name: Linnaeus named this taxon in 1753. *Selago* is an ancient name applied in pre-Linnaean times to a small firlike plant or species of *Lycopodium*. This plant is also commonly known as "cliff" or "mountain" clubmoss.

Horizontal Stem: Soon becoming erect. Rarely creeping.

Erect Stems: Equal length, 10 to 15 cm long. Forking from base to form clump. Densely tufted. Dark green, evergreen. Fertile and sterile leaves alternating in groups. but same size without distinct zones.

Sterile Leaves: 0.3 to 1 cm long. Linear, broadest at base. Margins entire.

Sporophylls: Same as sterile leaves. Sporangia borne in axils. Gemmae in upper leaves produce extensive vegetative reproduction. Gemma lobes narrow with sharp points.

Habitat: Mountainous regions in prevailing acidic soils. On cliffs and ledges and in crevices in sheltered or exposed situations.

Range: There are four varieties of *L. selago*. The Georgia variety is var. *appressum* Desv., which occurs in the mountainous areas of New England and in the southern Appalachians. In Georgia, it has been found only in Rabun County.

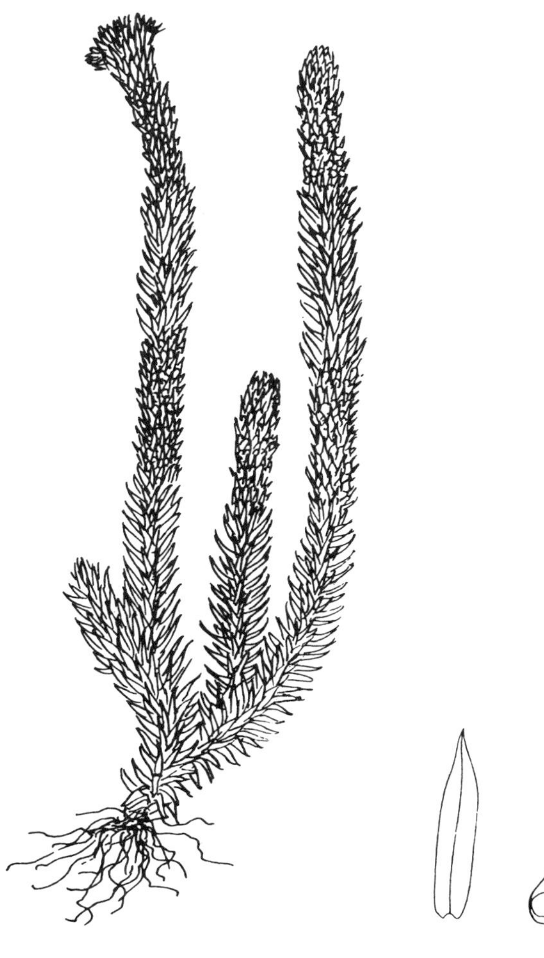

Foxtail Clubmoss

Lycopodium alopecuroides L.

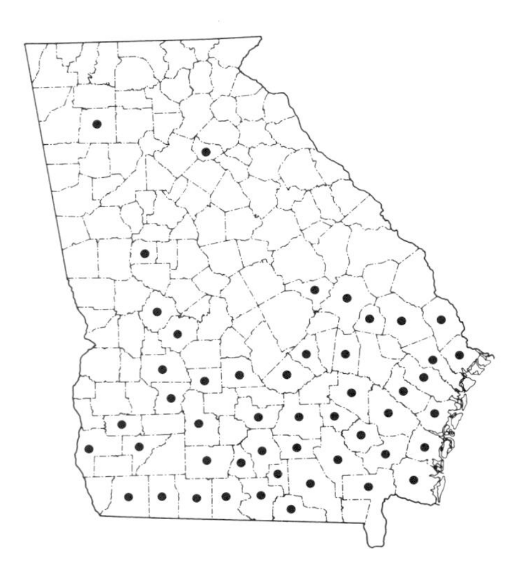

Name: The foxtail clubmoss is an American species named by Linnaeus in 1753. The epithet means "like *Alopecurus,*" a genus of grasses, and is derived from the Greek meaning "foxtail," in allusion to the brushlike fertile spike of the clubmoss and the spikelike fruiting heads of the grass.

Horizontal Stem: Long-creeping, strongly arching, rooting at intervals.

Erect Stems: 5 to 8 in number. Upright, 15 to 25 cm long. Usually deciduous, although somewhat evergreen and persistent in Georgia.

Sterile Leaves: In many rows. Narrowly linear with long, sharp points and conspicuous, marginal teeth. Wide-spreading.

Strobili: Bushy. 2.5 to 6 cm long. 1.4 to 1.6 cm wide.

Sporophylls: Longer and narrower than sterile leaves. Wide-spreading at essentially 90 degree angle to axis of strobilus.

Habitat: Wet, sandy meadows and pinelands, sphagnous meadows and depressions. Strongly acidic soil.

Range: Throughout the Gulf and Atlantic Coastal Plains of the southeastern United States from Texas to Florida and northward along the coast to Nantucket. In Georgia, throughout the Lower Coastal Plain, in much of the Upper Coastal Plain, and in a few counties of the Piedmont.

Hybrid Bog Clubmoss

Lycopodium alopecuroides × prostratum

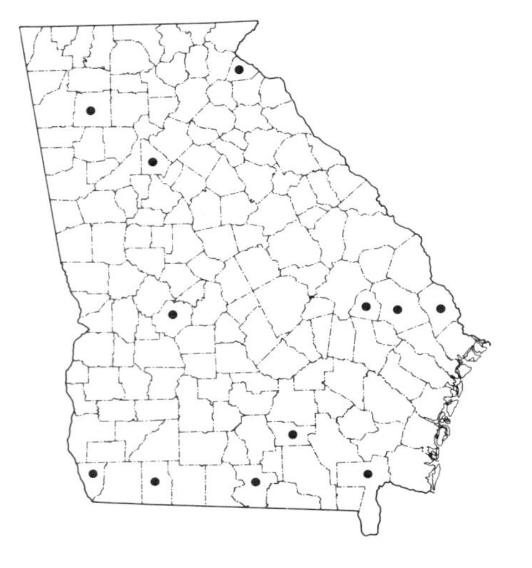

Name: The names of the bog clubmoss hybrids are denoted by the standard method of connecting the parents in alphabetical order with a crossbreed sign.

Horizontal Stem: Long-creeping, arching moderately, rooting along creeping portion.

Erect Stems: 1 to 4 stems. 15 to 30 cm long. Unbranched, deciduous.

Sterile Leaves: Long, slender, toothed. Those of horizontal stem wide-spreading, those of erect stem spreading to approximately 30 degree angle.

Strobili: Bushy. 3.0 to 8.0 cm long and 1.0 to 2.0 cm wide.

Sporophylls: Narrow, wide-spreading, toothed.

Habitat: Acidic sandy soils, wet ditches, wet piney meadows.

Range: Chiefly from North Carolina to Texas in the coastal plains. In Georgia, mostly in the Lower Coastal Plain.

Remarks: This hybrid is readily recognized by its bushy strobilus, moderate rather than strong arch in the horizontal stem, and angled leaves of the erect stem.

Pedunccles

appressum

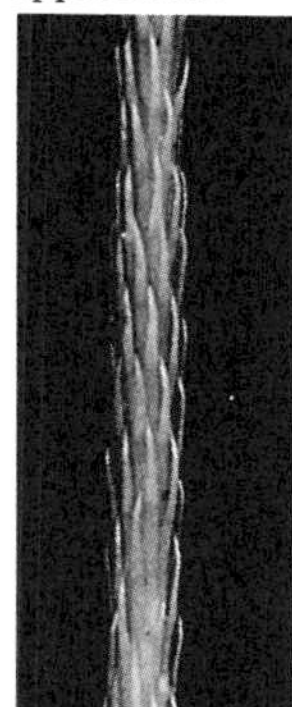

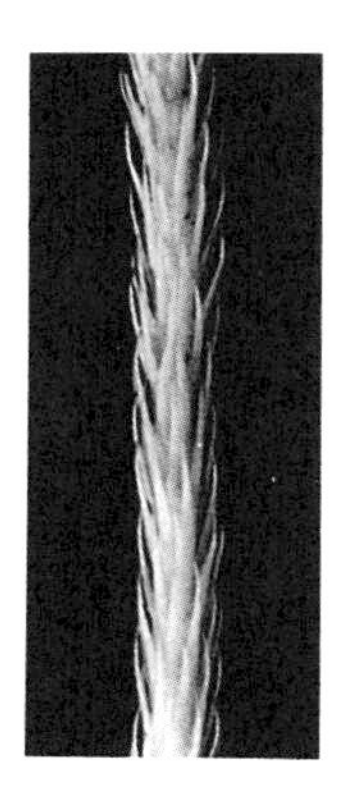

prostratum

alopecuroides

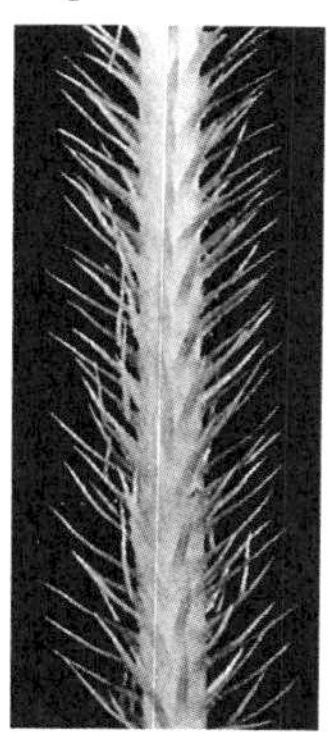

Hybrid Bog Clubmoss

Lycopodium alopecuroides × appressum

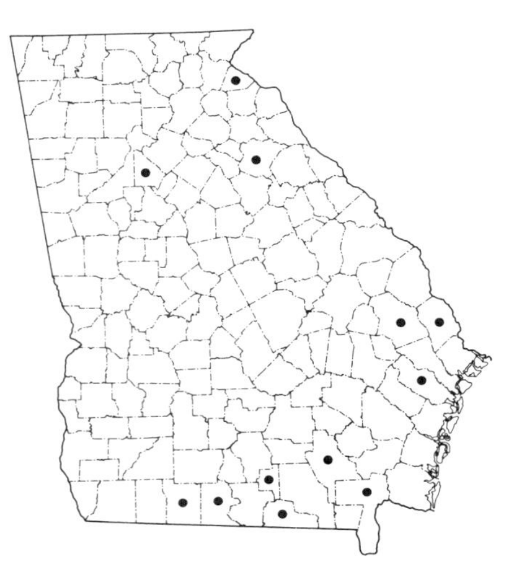

Name: As in the preceding account, the hybrid name is denoted by placing a crossbreed sign between the names of the parents. This plant is also known by the little-used name of *Lycopodium × copelandii* Eiger, which was published in 1956.

Horizontal Stem: Long-creeping, arching moderately, rooting along creeping portions.

Erect Stems: 4 to 8 stems. 15 to 30 cm long. Unbranched, deciduous.

Sterile Leaves: In many rows, narrowly linear, toothed on margins. Leaves on erect stems spreading at approximately 45 degrees to axis of stem.

Strobili: 2.5 to 6.0 cm long, and 0.4 to 1.2 cm wide. Somewhat bushy.

Sporophylls: Long and narrow, spreading at approximately 45 degree angle from axis of strobilus.

Habitat: Sandy ditches, wet sandy meadows, and wet pinelands. Acidic soils.

Range: Found throughout the overlapping portions of the parents' ranges. Numerous locations in Georgia, especially in the Lower Coastal Plain.

Remarks: This plant is probably the most common of the hybrids of the southern bog clubmosses because its parents more frequently grow near each other. It is distinguished by the intermediate spreading of the sporophylls and the moderate rather than strong arch in the horizontal stem.

Stems

appressum

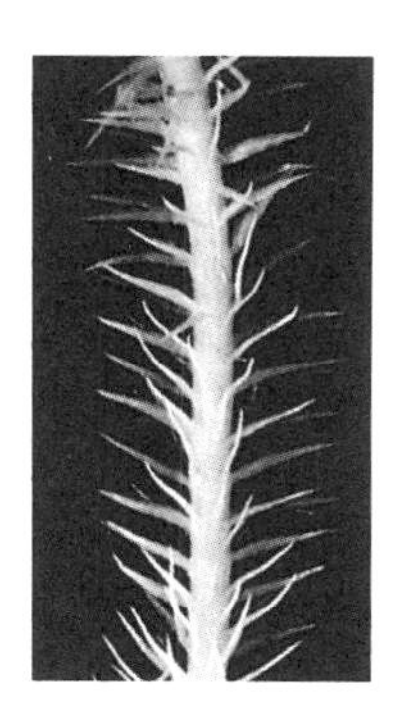

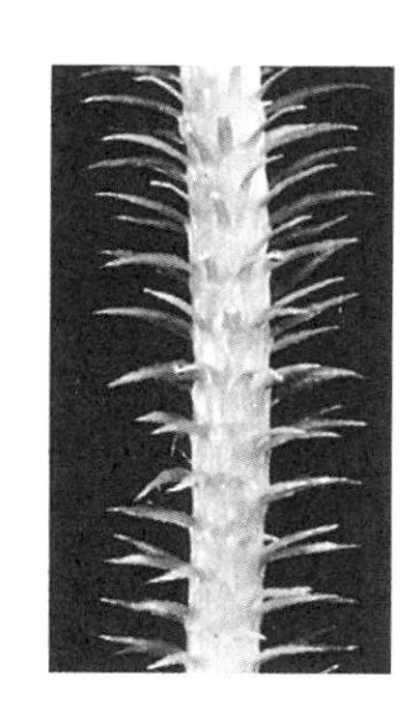

prostratum

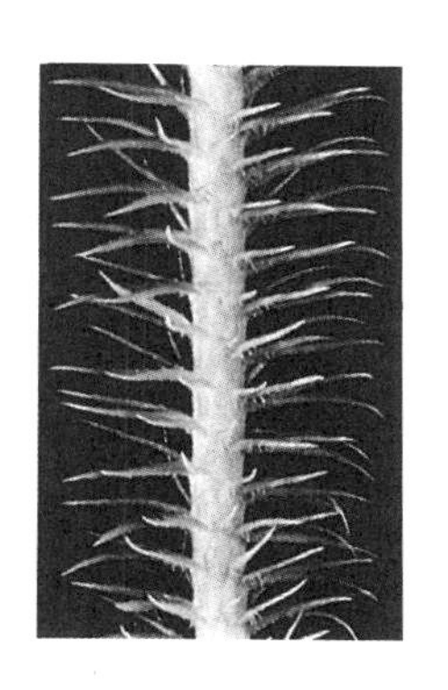

alopecuroides

Prostrate Clubmoss

Lycopodium prostratum Harper

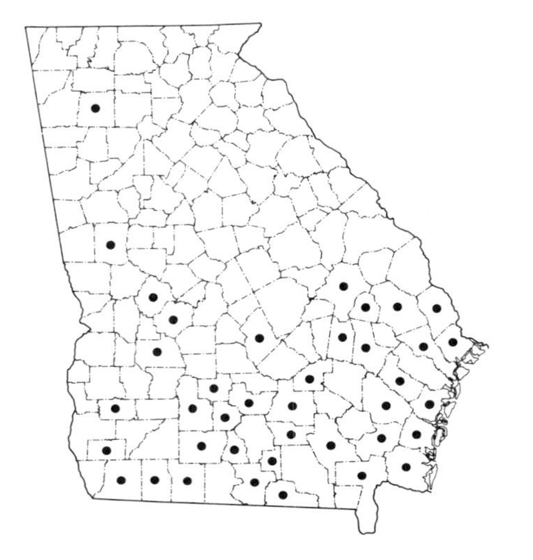

Name: In 1860 Chapman named this *Lycopodium inundatum* var. *pinnatum*. Lloyd and Underwood in 1900 raised it to specific rank, *Lycopodium pinnatum* ("featherlike in pattern"). Harper, in 1906, pointing out that this name was already in use, proposed *prostratum*, in reference to the prostrate sterile stem. It is also known as "southern clubmoss" and "feather-stem clubmoss."

Horizontal Stem: Prostrate, long-creeping, rooting essentially throughout.

Erect Stems: 15 to 30 cm long and 0.4 to 0.8 cm wide. Fertile branches 1 to 3, unbranched. Deciduous.

Sterile Leaves: On horizontal stem—long, slender, in many rows, spreading, strongly toothed on margins. On erect stems—slender, pointed, ascending, nearly appressed, toothed.

Strobili: 4 to 8 cm long and 1 to 2 cm wide. Bushy.

Sporophylls: Slightly larger than sterile leaves. Strongly toothed, wide-spreading.

Habitat: Acidic sandy soil, moist boggy depressions, pinelands and meadows.

Range: Throughout Atlantic and Gulf lowlands from North Carolina to Texas. In Georgia, common throughout the coastal plains, especially the Lower Coastal Plain.

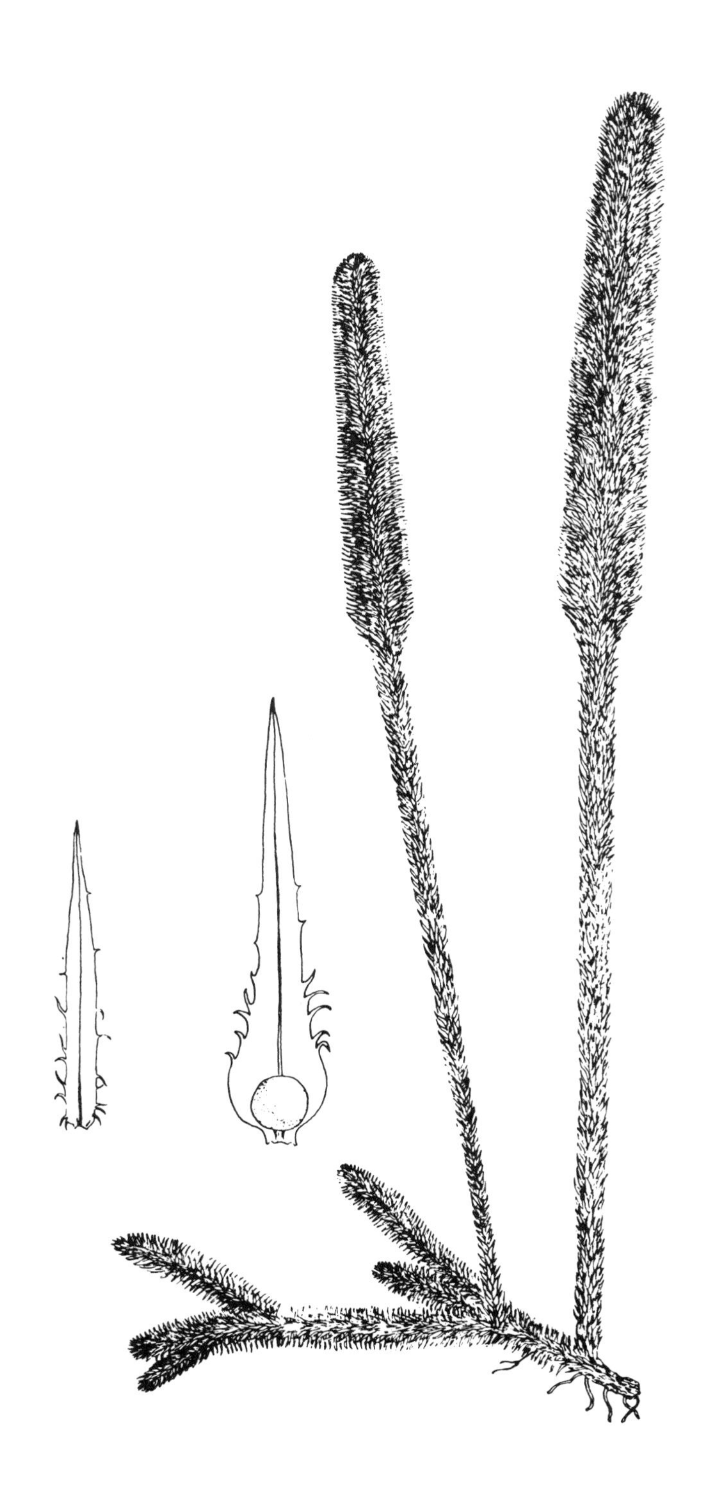

Hybrid Bog Clubmoss

Lycopodium appressum × prostratum

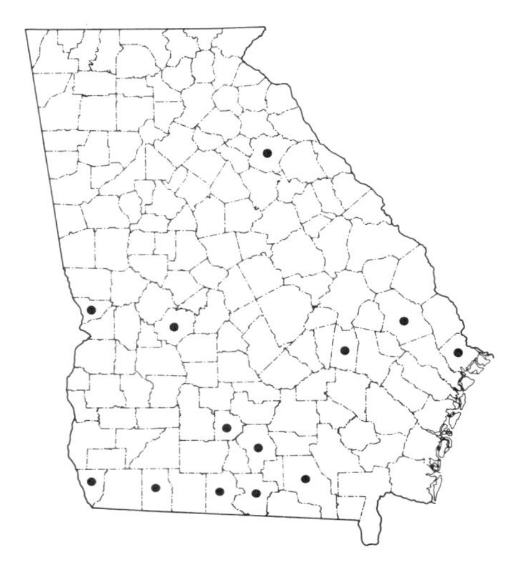

Name: As in the preceding account, the hybrid name is denoted by placing a crossbreed sign between the names of the parents. This plant is also known as *Lycopodium × brucei* (Cranfill) Lellinger, published in 1981 and 1984.

Horizontal Stem: Not arching, long-creeping, rooting essentially throughout.

Erect Stems: 1 to 6, unbranched. 15 to 30 cm long. Deciduous.

Sterile Leaves: In many rows, narrowly linear, toothed. Leaves of erect stems spreading to approximately 45 degree angle.

Strobili: 3 to 7 cm long and 0.5 to 1.3 cm wide. Somewhat bushy.

Sporophylls: Long and narrow, spreading to approximately 45 degree angle from axis of strobilus.

Habitat: Acidic sandy soils, sandy roadside ditches, wet piney meadows.

Range: North Carolina to Texas in the coastal plains. In Georgia, primarily in the Lower Coastal Plain.

Remarks: Readily distinguished by the intermediate spreading of the sporophylls and the lack of arching in the horizontal stem.

Strobili

appressum

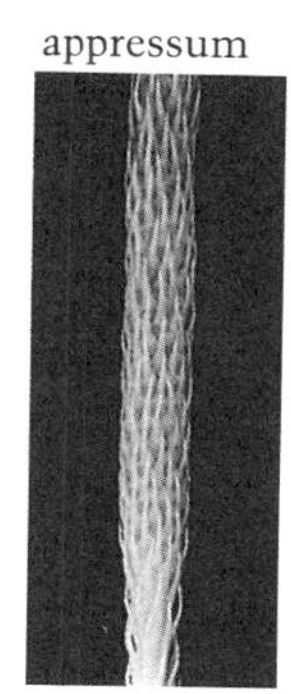

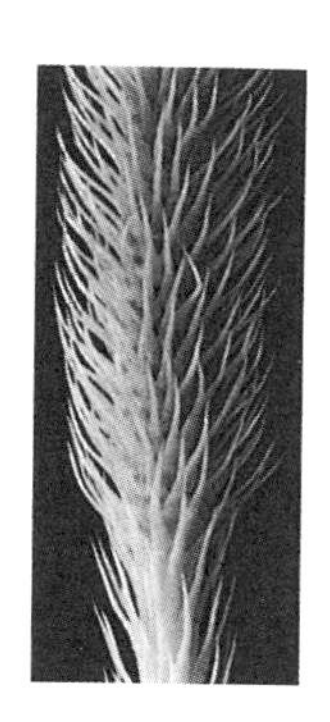

prostratum

alopecuroides

Southern Clubmoss

Lycopodium appressum (Chapm.) Lloyd and Underw.

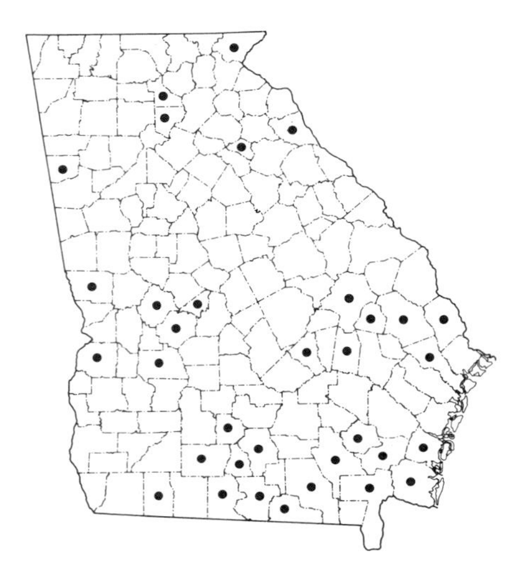

Name: A. W. Chapman, a nineteenth-century student of southern flora, proposed the name *Lycopodium inundatum* var. *appressum* in 1878. Lloyd and Underwood raised it to an independent species in 1900. The name *appressum* refers to the closely appressed spore-bearing leaves. The plant is also commonly called "slender clubmoss" and "appressed clubmoss," both terms being descriptive.

Horizontal Stem: Long-creeping, prostrate, rooting throughout. Leaves in many spiral rows, spreading and toothed; most partially turned upward. Those on lower side reduced.

Erect Stems: 4 to 15 stems. 15 to 30 cm long. Unbranched, deciduous.

Sterile Leaves: Long, slender, appressed to stem, sparingly or not toothed.

Strobili: 2.5 to 5.0 cm long and 0.3 to 0.6 cm wide. Long, slender.

Sporophylls: Long, slender, entire. Tightly appressed.

Habitat: Sphagnum bogs, meadows, wet sand barrens. Intensely acidic soil.

Range: In eastern North America from Newfoundland southward to Florida and westward to Texas and Arkansas; chiefly in the Gulf and Atlantic coastal lowlands. In numerous locations throughout Georgia, especially in the Lower Coastal Plain.

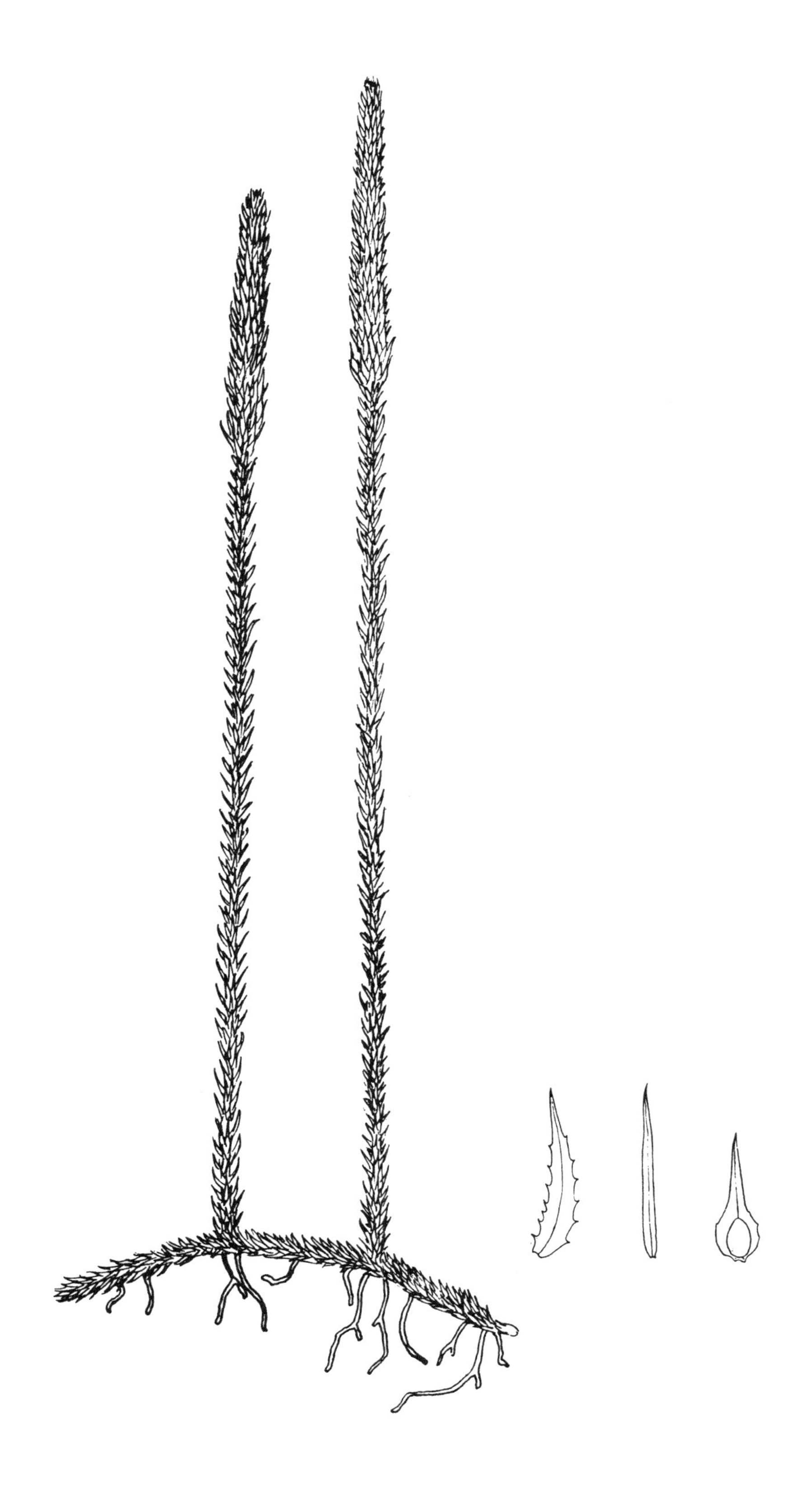

Carolina Clubmoss

Lycopodium carolinianum L.

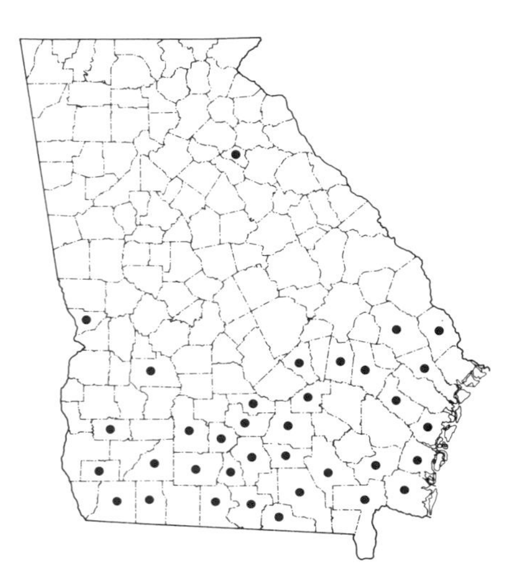

Name: Linnaeus named this taxon in 1753 from a plant that came to him from "Carolina." The plant is also commonly called "slender clubmoss."

Horizontal Stem: Prostrate, deciduous and short-creeping. Irregularly branched. Leaves entire, relatively broad.

Erect Stems: Very slender and sparsely leaved. 15 to 25 cm long.

Sterile Leaves: Narrow, scattered. Non-overlapping in pseudowhorls of three.

Strobili: Yellowish. 3 to 8 cm long and 2.5 to 4 mm wide.

Sporophylls: Broad, rhombic-ovate with long points. At first appressed, when mature the leaves bend outward, perpendicular to cone, giving it a spiky appearance.

Habitat: Sandy soil in moist, acidic pinelands and depressions. Often associated with other coastal lycopods.

Range: Tropics and subtropics of the Old and New worlds. Restricted in the United States to the eastern and southern coastal plains from Massachusetts southward to Florida and westward to Texas. In Georgia, in practically all counties of the Lower Coastal Plain and in a few counties of the Upper Coastal Plain.

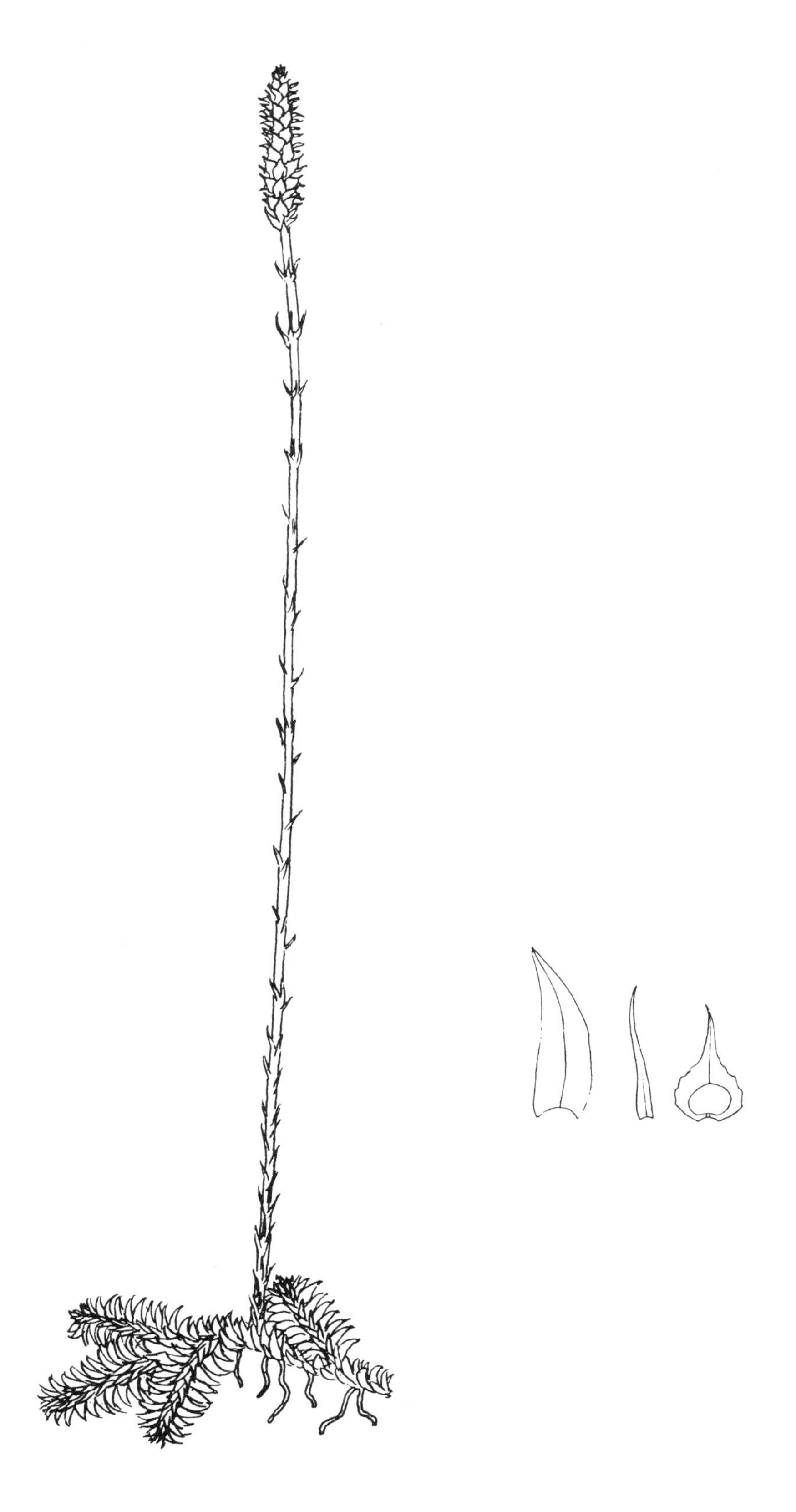

Nodding Clubmoss

Lycopodium cernuum L.

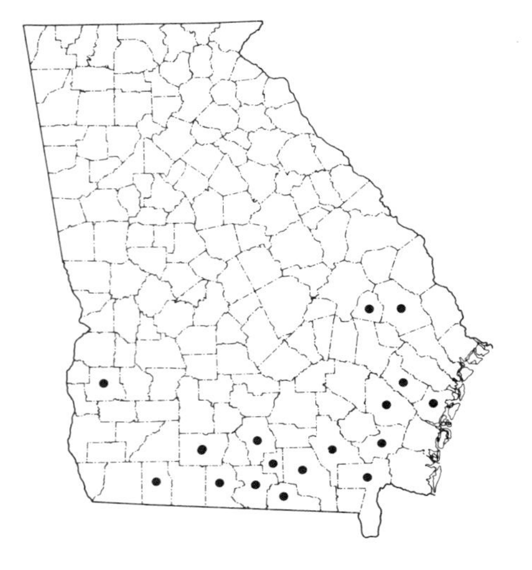

Name: Linnaeus described this taxon in 1753. It was originally found in the East Indies and subsequently in America. *Cernuum* means "nodding," in reference to the position of the fertile cones.

Horizontal Stem: Reclining, arching, or creeping. Occasionally sending out branches that root and send up new erect stems.

Erect Stems: 30 to 70 cm long. Much-branched. Christmas tree–like. Very leafy throughout.

Sterile Leaves: Slender, needlelike, of equal length. 0.3 to 0.5 cm long. Somewhat curved. Spirally arranged in many rows.

Strobili: 0.5 to 1 cm long and less than 2 mm wide. Numerous, sessile, compact. Drooping at tips of branches.

Sporophylls: Triangular-ovate. Coarse-toothed. Much broader than sterile leaves.

Habitat: Moist meadows, bog margins, stream banks, roadside ditches and cuts. Soil strongly acidic.

Range: Tropics of the Old and New worlds. In North America, from southern Georgia and Florida westward to Louisiana.

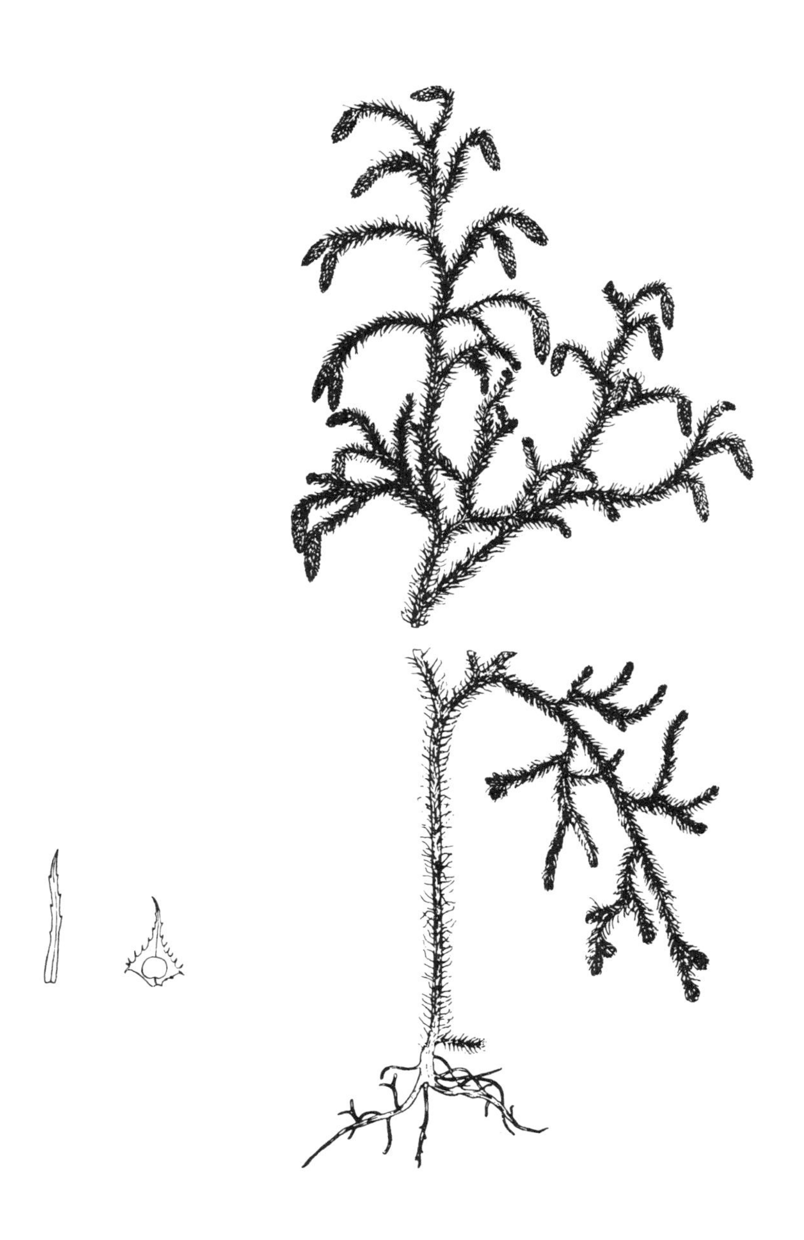

Ground Pine

Lycopodium clavatum L.

Name: Linnaeus named this taxon in 1753. The epithet means "club-shaped." This plant is also called "running clubmoss" and "staghorn clubmoss."

Horizontal Stem: Long-creeping on soil surface. Rooting throughout.

Erect Stems: 6 to 25 cm high. Branching several times, resembling deer antlers. Evergreen.

Sterile Leaves: In 10 rows, ascending. 0.35 to 0.75 cm long with long, hair-like tip. Long, tapering with a narrow base. Yellowish green.

Strobili: 3 to 7.5 cm long. Usually 2 or 3 on peduncles 7.5 to 15 cm long.

Sporophylls: Orbicular-deltoid with long tip.

Habitat: Acidic soil in open woods, grassy thickets, rocky barrens, and swamp margins.

Range: Common in northern North America, Europe, and Asia. Found in Fannin County, Georgia, in 1984.

CCS 85

Tree Clubmoss

Lycopodium obscurum L.

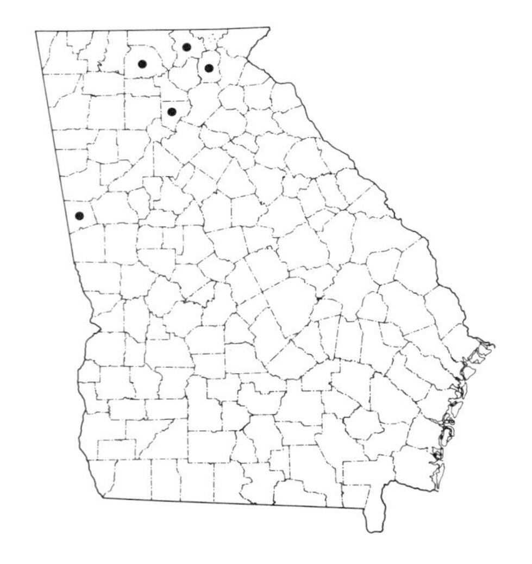

Name: Linnaeus in 1753 named this lycopod *obscurum,* or "hidden," referring to the deep-buried rhizome. The common name refers to its tree-like appearance. It is also called "ground pine."

Horizontal Stem: Creeping and branching. 3 to 6 cm deep in the ground.

Erect Stems: 15 to 30 cm long, with spreading, treelike branches, repeatedly branching upward. Give appearance of a tiny, thickly branched pine tree.

Sterile Leaves: Shining dark green, evergreen. Spreading, in 6 rows. Needlelike, narrow, 0.3 to 0.6 cm long, with pointed tips.

Strobili: Sessile at ends of branches. 3 to 6 cm long.

Sporophylls: Almost round with flat base and pointed tips.

Habitat: Acidic soils in moist or seemingly dry humus in rich deciduous and coniferous woods.

Range: Eastern North America from Canada westward to Minnesota and southward to Alabama and Georgia. In Georgia, in several northern counties.

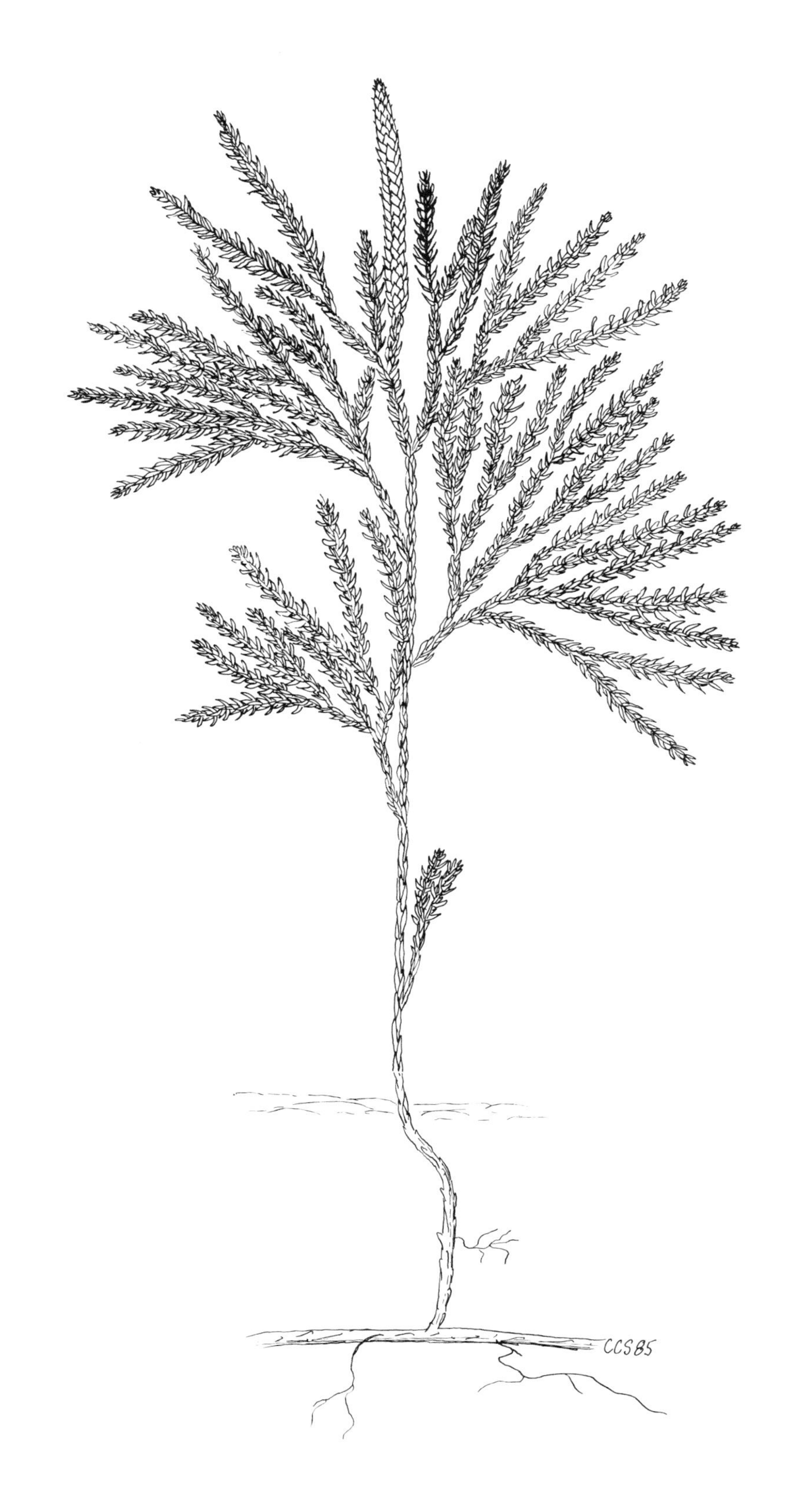
CCS85

Ground Cedar

Lycopodium tristachyum Pursh

Name: The American taxonomist Frederick Pursh named this lycopod in 1814. The species name means "three-spiked," so named because there are, on the average, three peduncles with strobili.

Horizontal Stem: Long-creeping. Deep underground.

Erect Stems: 15 to 30 cm long. Blue-green, evergreen. Branches about 1.5 mm broad, repeatedly branching. Lower branches fan-shaped, upper flat-topped. Annual constrictions from previous year's growth conspicuous.

Sterile leaves: In 4 rows, all similar in size and shape. Lance-shaped, sharp-pointed, fused together for about one-half of length.

Strobili: Naked peduncle, 5 to 10 cm long, twice-forked to bear four cones. Cones short-stalked, cylindrical, 1.5 to 3 cm long.

Sporophylls: Round with flat base and pointed tip.

Habitat: Acidic soils in deciduous and coniferous open woods, sandy meadows, and rocky barrens.

Range: Frequent in Europe and northeastern North America from Newfoundland westward to Minnesota and Missouri and southward barely to northern Georgia and Alabama. In Georgia, only in Habersham County.

Remarks: *L. tristachyum* hybridizes with some of the other clubmosses, most commonly with *L. digitatum* to form *L.* × *habereri* House.

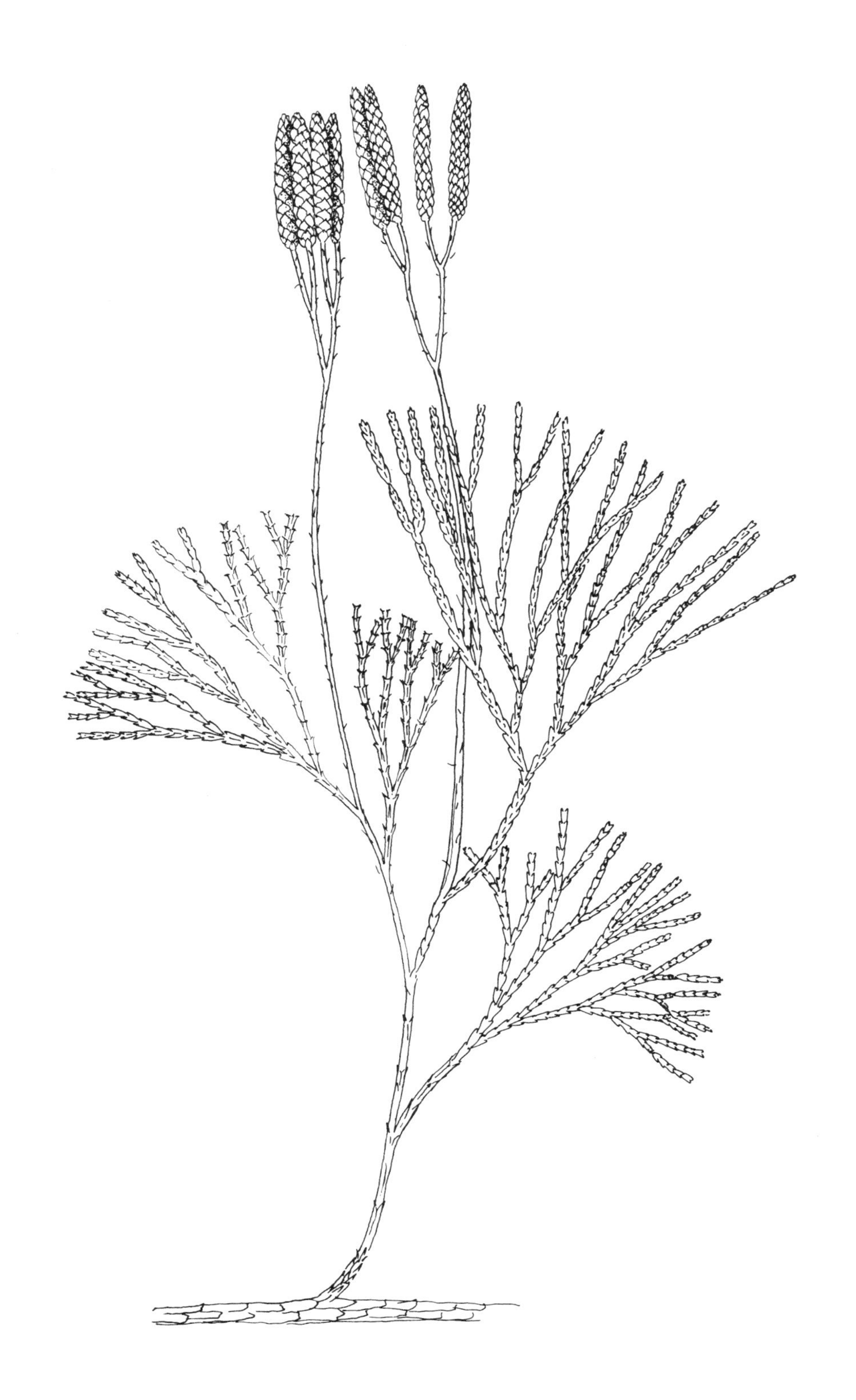

Running Ground Pine

Lycopodium digitatum A. Br.

Name: In 1901 M. L. Fernald recognized a common eastern taxon as *Lycopodium complanatum* var. *flabelliforme,* which Blanchard in 1911 raised to species rank. Recently, in 1979, Alexander Braun's earlier name (1848), *Lycopodium digitatum,* was rediscovered. *Digitatum* means "with branchlets radiating from the tips of the leaf stalks like long fingers." It is also commonly called "running cedar."

Horizontal Stems: Long-creeping, sometimes up to 100 or 200 cm. On soil surface or under leaf litter. Leaves few, distant, stubby.

Erect Stems: 12 to 25 cm long. Evergreen. Repeatedly branched in flattened, fanlike patterns.

Sterile Leaves: In four rows. Fused most of length to stem. Side leaves larger, broader, and more spreading than those on top and bottom of stem.

Strobili: On 4 to 10 cm long, naked peduncles. Usually forking twice and bearing four strobili 2 to 5 cm long, commonly with sterile tip.

Sporophylls: Yellow when mature. Orbicular-deltoid with pointed tips. Edges rough.

Habitat: Dry or barely moist, open or semiopen deciduous or coniferous woods, thickets, and grassy slopes. Acidic, humus-rich soils.

Range: Common in eastern North America from Canada westward to Minnesota and Kentucky and southward to northern Alabama and Georgia. In Georgia, found in the Blue Ridge and throughout the Piedmont, but nowhere in the Coastal Plain.

Remarks: Some Georgia specimens were identified as *L. complanatum,* a name at one time erroneously applied to *L. digitatum. L. complanatum* has not been found in Georgia. This latter plant has occasionally buried rhizomes; the branches grow in an irregular, not fanlike pattern; the annual constrictions from the previous year's growth are conspicuous; the strobili are in groups of from 1 to 4; the cones are only 1 to 3 cm long and lack the sterile tip.

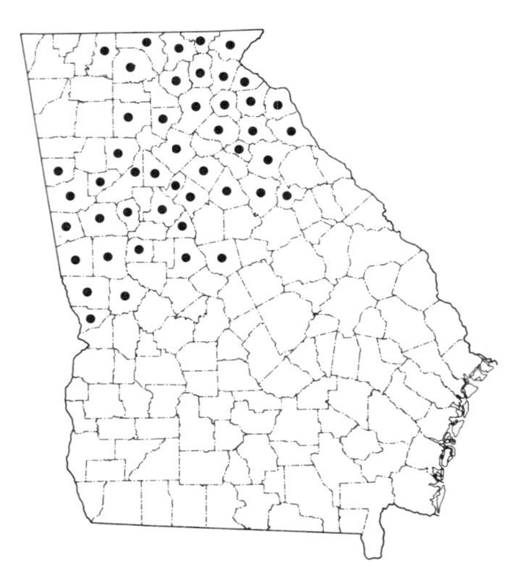

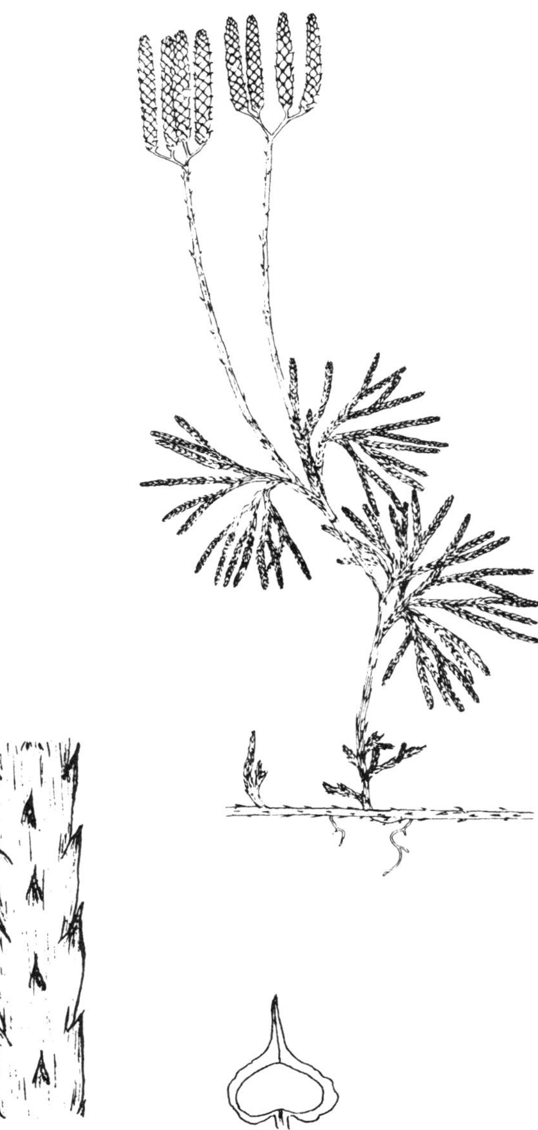

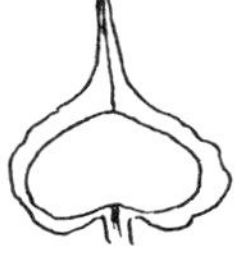

Blue Spikemoss

Selaginella uncinata (Desv.) Spring

Name: This Asian spikemoss was named *Lycopodium uncinatum* by Desvaux in 1814 and was transferred to *Selaginella* by Anton Spring in 1850. The epithet means "hooked."

Stems: Slender, weak, straw-colored. Creeping with flat, frondlike branches. Primary branches 3 to 5 cm long, ultimate branches less than 1 cm.

Leaves: Bluish green. 4 rows. Pale, shiny margins minutely toothed. Lateral ovate-oblong with short tips. Midvein prominent, especially upper half. Median leaves on lower side of branches overlapping, obliquely oblong with constricted sharp tips.

Strobili: Distinctly 4-sided. Sporophylls ovate-lanceolate, crowded, sharp-heeled, and ascending.

Habitat: Moist woods. Shaded and damp situations.

Range: Native to China. Naturalized from Florida to Louisiana. In Georgia, found in the late 1970s in Decatur County.

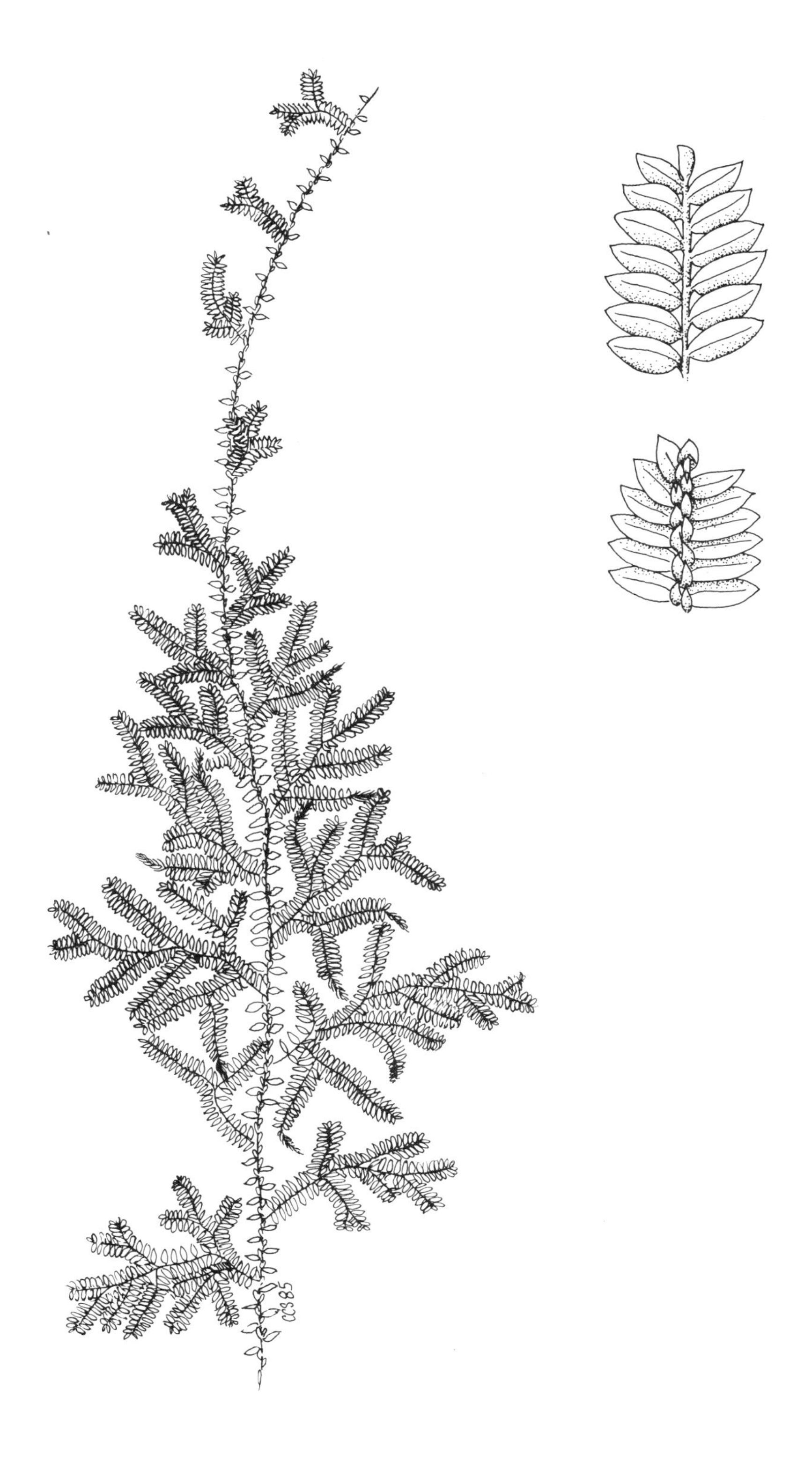

Mat Spikemoss

Selaginella kraussiana A. Br.

Name: Braun named this plant in 1860 in honor of Herr Kraus.

Stems: Prostrate, wide-creeping, 5 to 10 cm long. Branching.

Leaves: Green. Margins finely toothed. 4 rows. Lateral leaves oval. Median leaves smaller, linear, appressed with tip pointing forward.

Strobili: 2 to 3 cm long. 4-sided.

Habitat: Damp meadows and margins of ponds and streams.

Range: Commonly cultivated and sometimes escaped in the southeastern states. Found in Georgia only in Muscogee County.

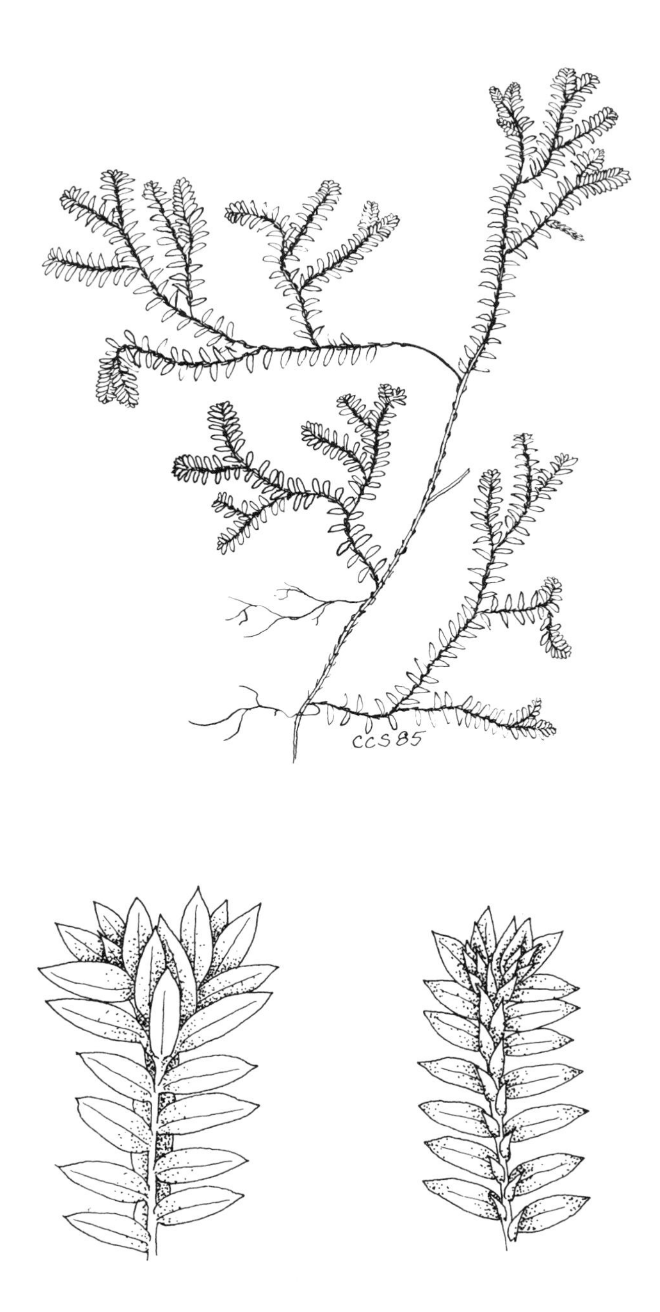
CCS 85

Meadow Spikemoss

Selaginella apoda (L.) Spring

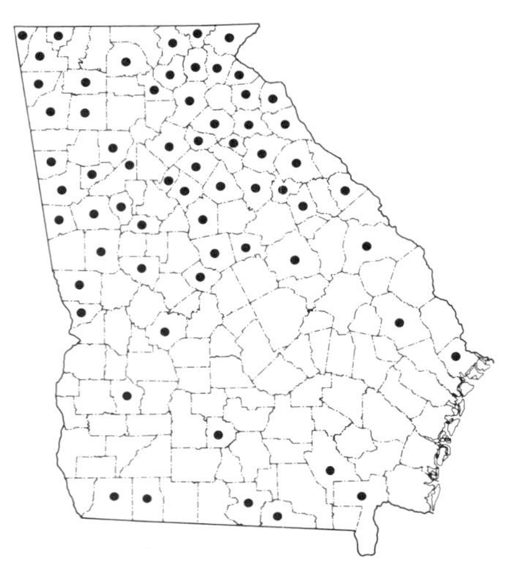

Name: Linnaeus in 1753 described *Lycopodium apodum* and in 1840 the Belgian taxonomist Anton Spring transferred it to *Selaginella.* The epithet means "footless," referring to the lack of heavy, creeping rootstock.

Stems: Creeping close to ground. Spreading and branching to form large, flat, dense mats. Slender roots emerge all along stem.

Leaves: Light green. Evergreen. 4 rows. Lateral leaves oval, margins finely toothed. Medial leaves smaller, narrowly oval or lanceolate with pointed tips, appressed and pointing forward.

Strobili: 1 to 2 cm long at ends of branches. 4-angled with sporophylls tightly bunched.

Habitat: Moist places in meadows, along stream banks, about springs, on ledges under waterfalls. Often in partial shade and best in somewhat calcareous soil.

Range: Widely distributed throughout the eastern United States. Found throughout Georgia, most commonly in the Piedmont and Blue Ridge.

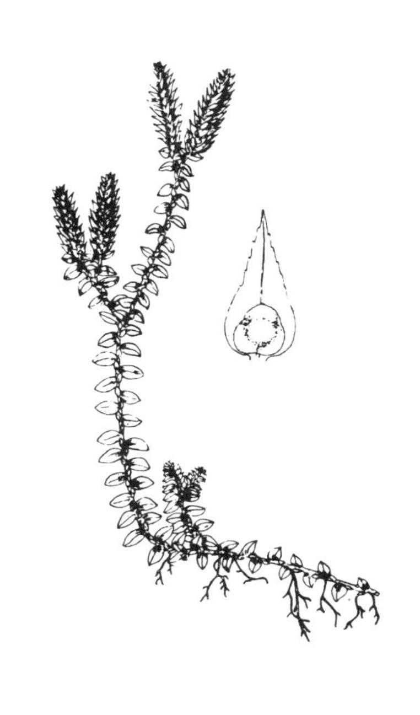

Gulf Spikemoss

Selaginella ludoviciana A. Br.

Name: Alexander Braun described this plant in 1860, the specific name meaning "of Louisiana," where it was reportedly first discovered about 1850.

Stems: Creeping at base. Roots at nodes. Shoots growing erect to 20 cm long.

Leaves: Bright green. Margins finely serrate and distinctly pale or clear. 4 rows. Lateral lanceolate, pointed and spaced. Median appressed, smaller, ovate-obtuse with tip pointed forward.

Strobili: 1 to 1.5 cm long. 4-angled. Sporophylls ovate-lanceolate, acuminate, minutely toothed above base.

Habitat: Circumneutral soil in swamp margins and moist, grassy openings in pinelands.

Range: The Gulf states from northern Florida to eastern Texas. In Georgia, only in a few counties in the southwestern part of the state.

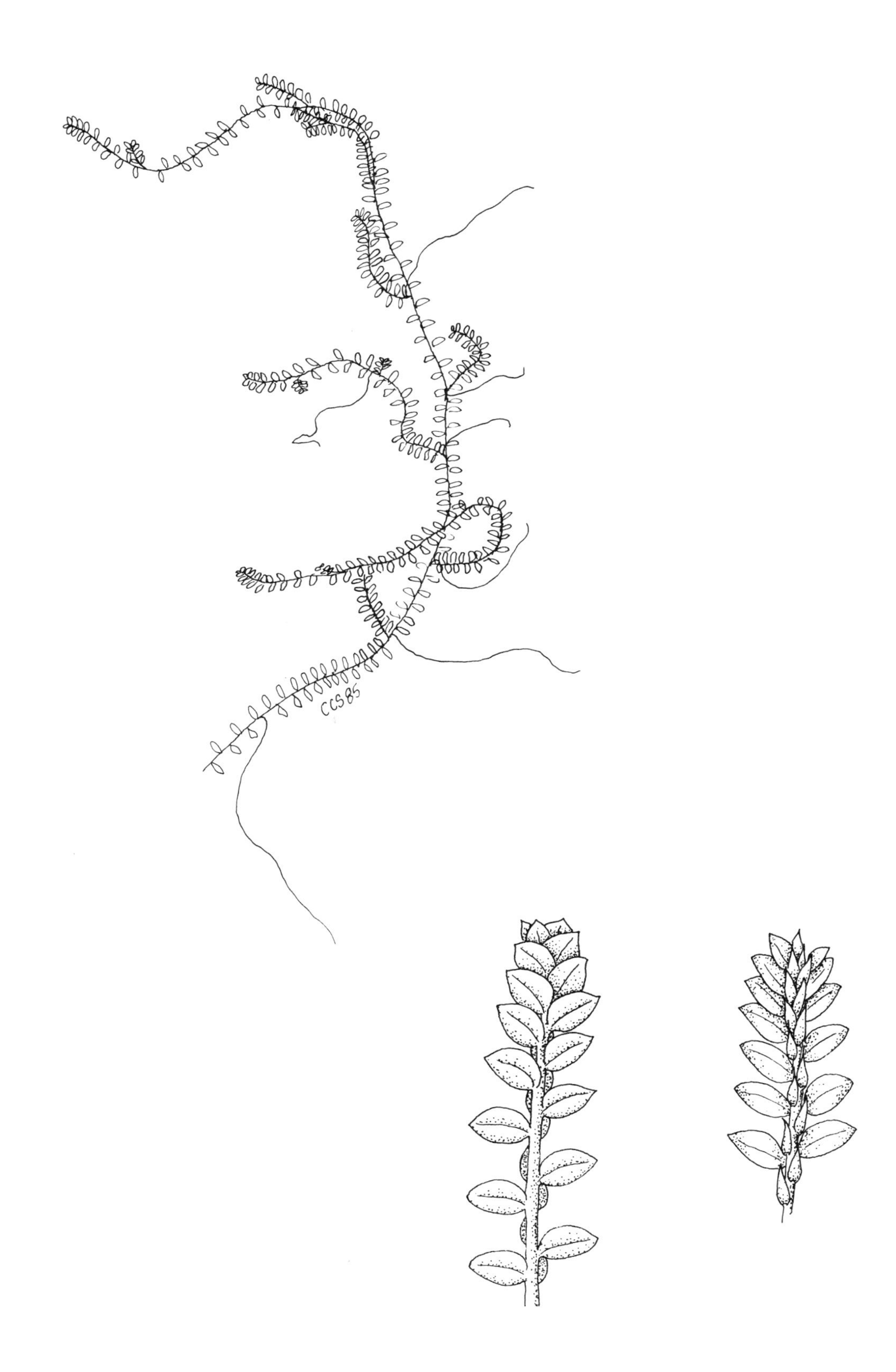
CCS85

Braun's Spikemoss

Selaginella braunii Baker

Name: The British pteridologist John Baker named this plant in 1867 in honor of the German taxonomist Alexander Braun.

Stems: Erect or ascending to form frondlike plants 20 to 45 cm long and 10 to 15 cm wide. Covered with short, pointed hairs, scarce at base becoming dense at apex. Upper three-fourths of plant much-branched with deltoid or ovate outline. Primary branches erect-spreading, widely spaced, alternate, triangular, with lower half broader than upper. Secondary branches deltoid, alternate, spaced. Erect stems woody, straw-colored. Branch stems light green and hairy. Rhizome creeping, hairy, subterranean.

Leaves: Gray-green. On erect stem in 4 rows. Small, ascending, oblong with pointed tips. Distant at base becoming closer at apex. Branch leaves firm in texture, in 4 rows in two planes. Lateral leaves ovate, often rolled with margins underneath. Dorsal leaves smaller, oblong, lower sides unequal, with white elongate tips. Leaves distant becoming confluent on ultimate branches.

Strobili: Short, slender, 4-sided.

Habitat: Wet woods, swamps, and meadows.

Range: A native of eastern Asia cultivated in a few gardens and rarely escaped in the southeastern United States. In Georgia, only in Camden County.

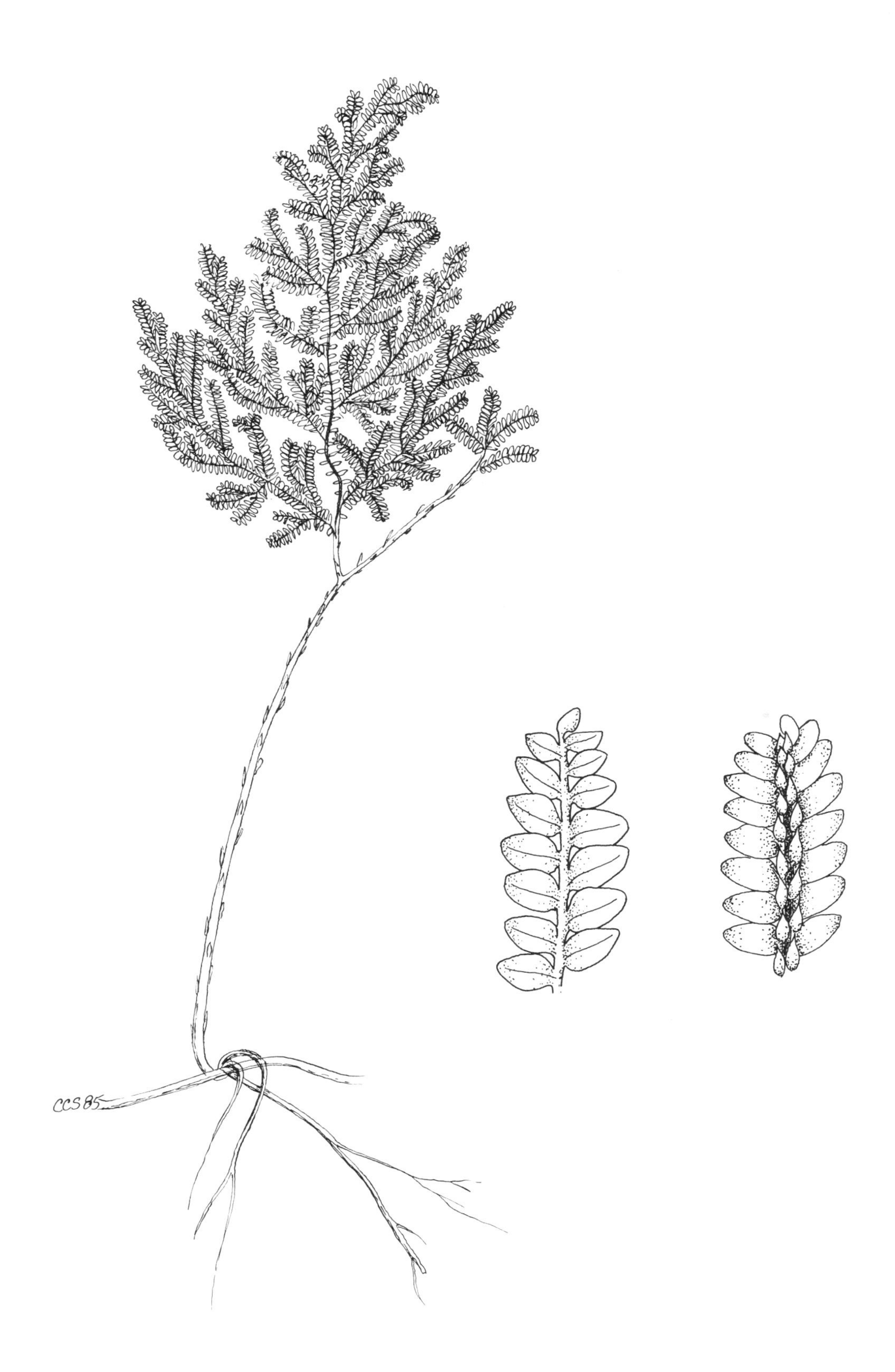
CCS 85

Sand Spikemoss

Selaginella arenicola Underw.

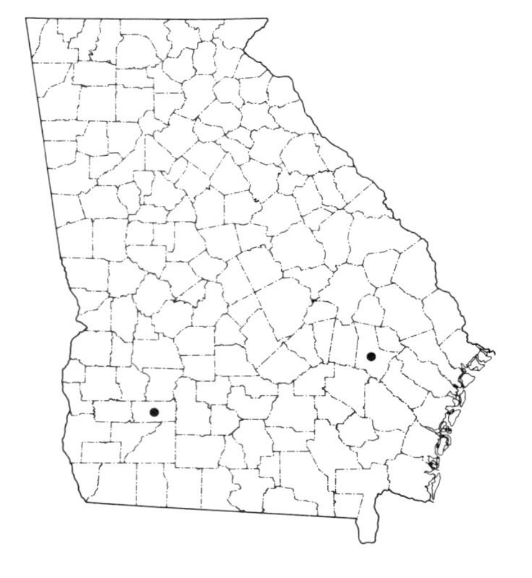

Name: Originally found in Florida, this plant was named by L. M. Underwood in 1898. The Latin epithet means "growing in sand," in reference to its habitat.

Stems: Slender, erect, forming tufts 4.5 to 6 cm long. Including leaves, 1.0 mm or less in diameter when moist. Rooting only near base.

Leaves: All alike, usually in 6 rows. Long-oblong, 1 to 1.5 mm long, with long decurrent base, with tiny hairs at base. Overlapping. Furrowed longitudinally on outer side. Marginal hairs present, medial ones confined to base of furrow. Setae toothed and straight, colorless, without a red band at base.

Strobili: Numerous, erect, 1 to 2 cm long. Slender, 4-sided, conspicuously larger in diameter than vegetative stems. Sporophylls linear-deltoid, with auriculate base, short tip, and hairy margins. Sulcate.

Habitat: Sand hills and barrens, pine woods, and gravelly granite slopes. Soil strongly acidic.

Range: Peninsular Florida and southern Georgia.

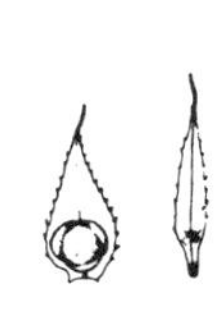

Spiny Spikemoss

Selaginella acanthonota Underw.

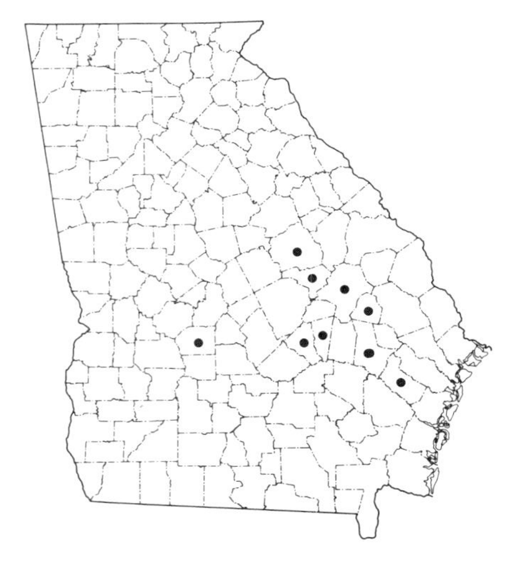

Name: This species was described by L. M. Underwood in 1902. The name means "with a row of spines," referring to the minute, stiff hairs on the dorsal surface of the leaf along the furrow.

Stems: Slender (but not as slender as *S. riddellii* or *S. arenicola*), suberect to creeping, forming tufts 3 to 6 cm long. Including leaves, 1.5 to 2.3 mm wide when moist. Densely and repeatedly branched. Rooting from middle of stem or above.

Leaves: Ascending to appressed, overlapping over half of length. Narrowly deltoid from short, nondecurrent, obdeltoid base. Dorsally sulcate, with bristle point, often with reddish band at base. Hairy on margins and sparsely so along entire length of dorsal furrow.

Strobili: 1 to 3 cm long, slender, 4-sided. Sporophylls lanceolate-deltoid with auriculate base. Hairy on margins and slightly hairy on dorsal furrow.

Habitat: Sand barrens and dunes, and gravelly granite slopes. Soil strongly acidic.

Range: Along coastal plains from south-central Georgia to North Carolina.

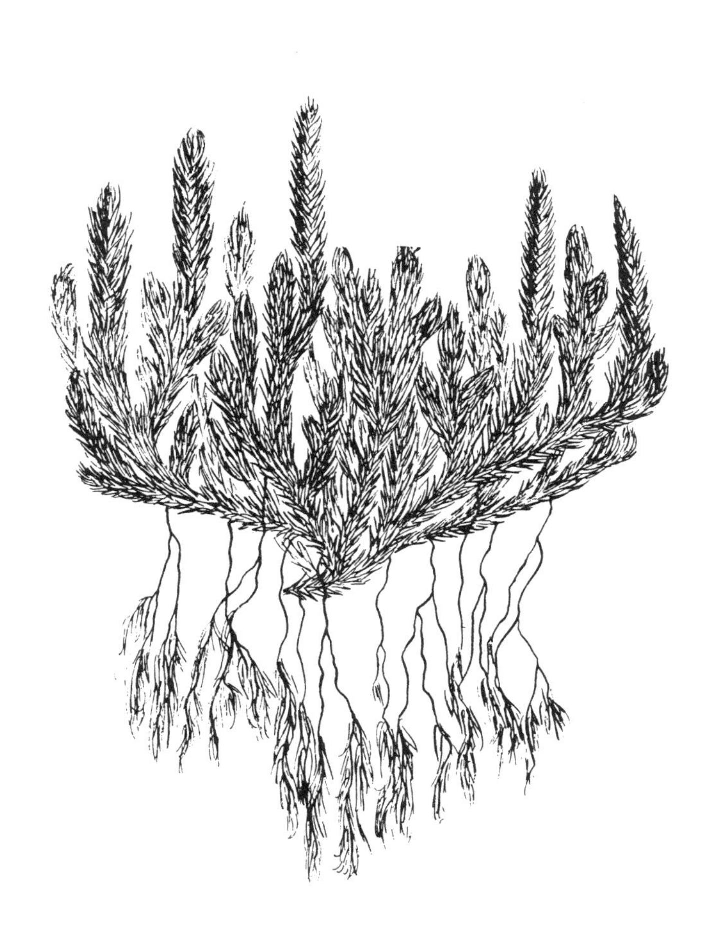

Riddell's Spikemoss

Selaginella riddellii Van. Eselt.

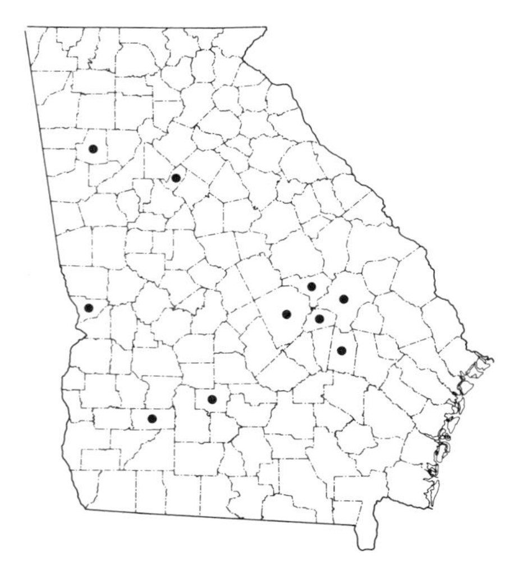

Name: In 1918 Van Eseltine described this species and named it in honor of J. L. Riddell.

Stems: Slender, erect, forming tufts 4 to 9 cm tall. Including leaves, greater than 1.0 mm in diameter when moist. Plant base well buried. Rhizome short, rooting only near base. 2 to 4 primary branches, 2.5 to 6 cm long, with few ascending branchlets.

Leaves: Ascending to appressed, overlapping, to 1.5 mm long. Long-oblong, with short, deltoid, decurrent base. Dorsally sulcate, with bristle point, marginal fine hairs, median ones confined to base of furrow.

Strobili: 1 to 2.5 cm long, slender, 4-sided, about the same diameter as the vegetative stems. Remains green after leaves below wither. Sporophylls elongate-triangular, up to 1.0 mm long. Minutely hairy margins. Setae similar to those of stem leaves.

Habitat: Gravelly granite slopes and outcrops; occasionally sand barrens. Soil strongly acidic.

Range: Inland from Texas to western Georgia. Found in scattered Georgia counties.

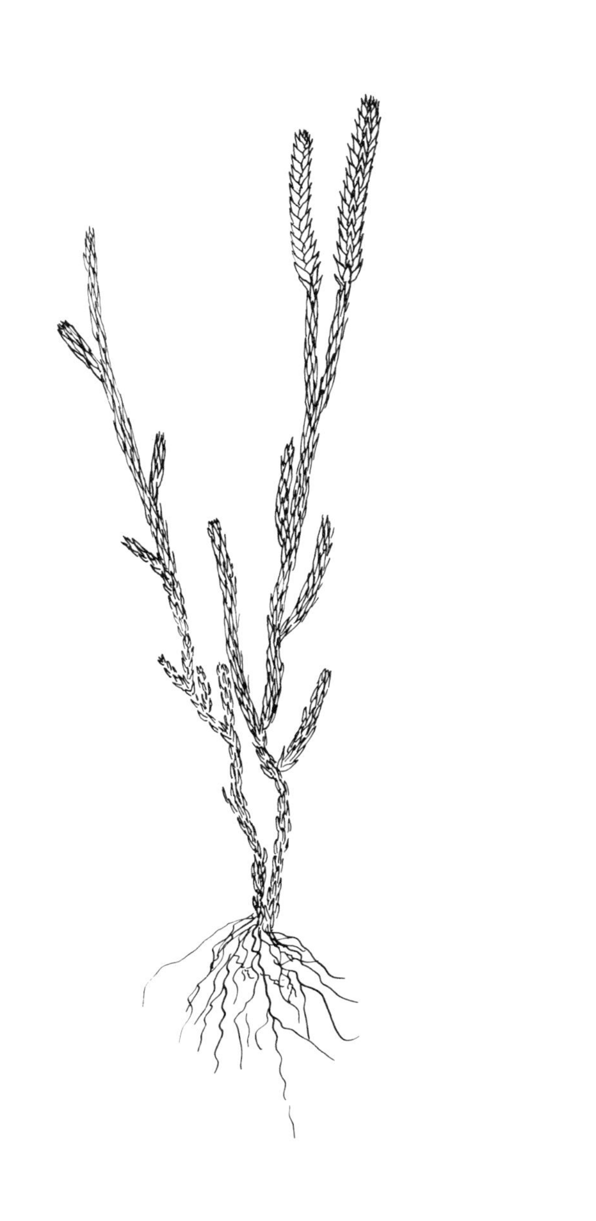

Twisted-Hair Spikemoss

Selaginella tortipila A. Br.

Name: First collected in 1839, this species was confused with *S. rupestris* until Alexander Braun described it as a separate species in 1865. The common name and epithet refer to the twisted setae.

Stems: Erect, forming dense tufts or mats 8 to 15 cm long. Stems nearly or quite prostrate around margins of clumps and throughout heavily shaded clumps.

Leaves: All alike. Closely overlapping, thick and plump. Furrowed on back for short distance but not at base. About 12 rows spirally arranged. Linear, tipped with much-bent and twisted white bristle.

Strobili: 4-sided, usually not over 5 mm long. Sporophylls ovate with twisted bristle tip. Margins smooth or with short teeth.

Habitat: Bare slopes of granite or granite-gneiss rocks, ledges, and shallow depressions. Often at high altitudes. Soil acidic.

Range: Frequent, but geographically restricted to eastern Tennessee, western North and South Carolina, and northeastern Georgia.

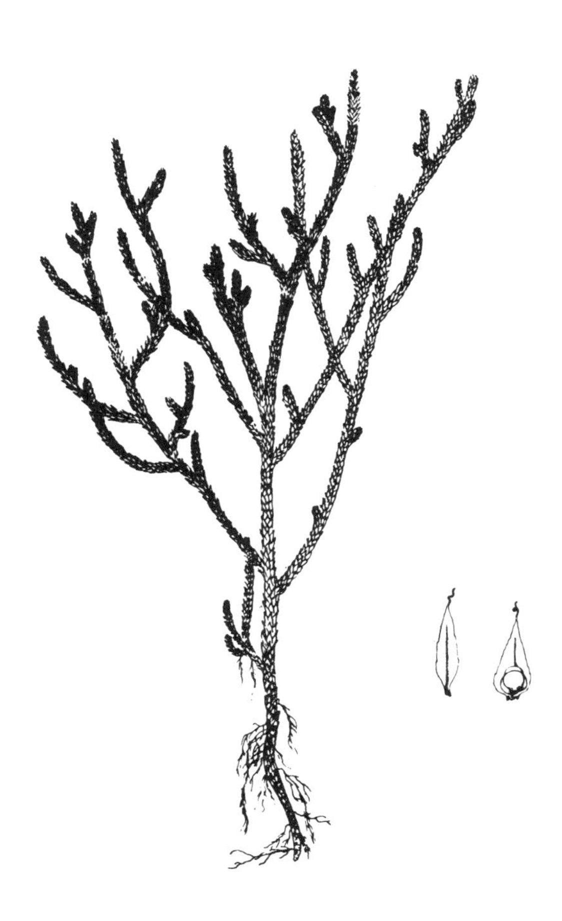

Rock Spikemoss

Selaginella rupestris (L.) Spring

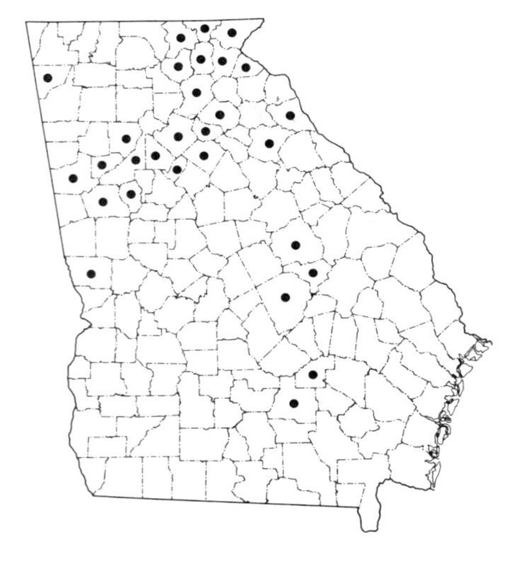

Name: In 1753 Linnaeus named this species *Lycopodium rupestre* from both Old and New World plants. Spring in 1838 transferred it to *Selaginella.* The epithet, meaning "of rocks," is appropriate because of its habitat.

Stems: Creeping and spreading, sending up ascending branches 3 to 7 cm long, forming tufts or mats. Roots slender, branching from tufted stems.

Leaves: Gray green, evergreen. Main stem and branches completely covered by several rows of appressed and overlapping leaves. Long and narrow, convex and deeply grooved with marginal hairs and long, firm, white terminal bristles.

Strobili: 1 to 2 cm long. Distinctly 4-sided. Sporophylls ovate with pointed apex and short bristle tips.

Habitat: Gravelly humus over sandstone, granite, and similar hard rocks. Full sun or part shade. Soil acidic or subacidic.

Range: The eastern United States from Nova Scotia to Minnesota and southward to Oklahoma and northern Alabama and Georgia. Widespread throughout northern Georgia.

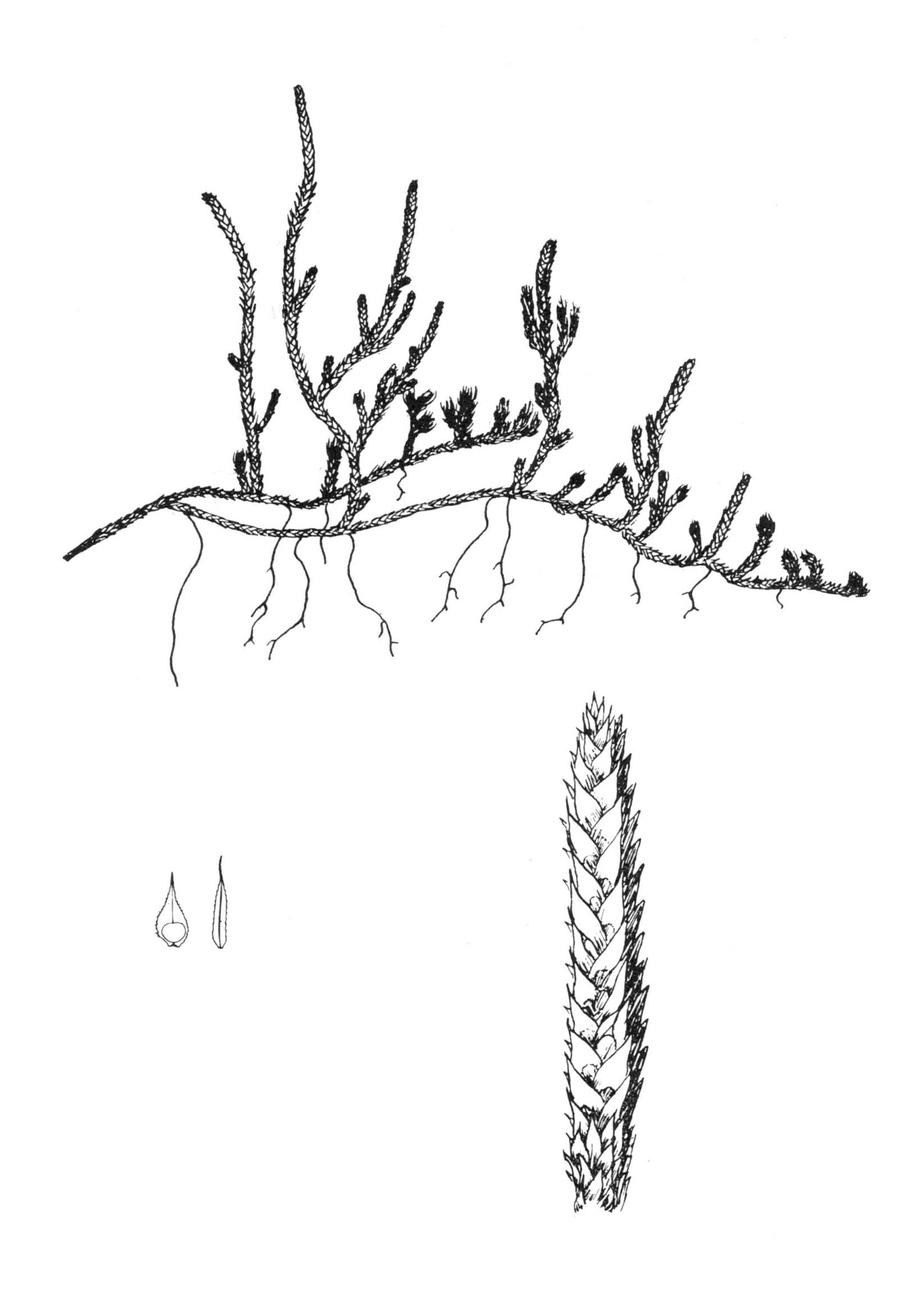

Engelmann's Quillwort

Isoetes engelmannii A. Br.

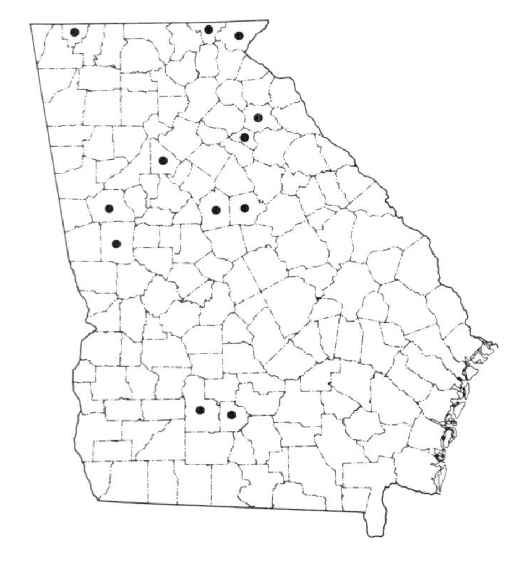

Name: This species was discovered near St. Louis by Engelmann in 1842 and was named in his honor four years later by Alexander Braun.

Roots: Monomorphic, forked. Corm 2-lobed.

Leaves: 25 to 60 cm long. 15 to 100 on each plant. Flaccid when submerged, stiffer when exposed. Stomata and peripheral strands present.

Sporangia: Elliptical. 6 to 12 mm long. No pigmented cells. Velum covers one-third to two-thirds.

Spores: Mature in late summer or autumn. Megaspores, 400 to 600 microns in diameter, light cream–colored or white, reticulate. Microspores, 20 to 30 microns long, light-colored, echinate or almost smooth.

Habitat: Amphibious. Wet ditches, running or standing shallow fresh water.

Range: Most common of the southeastern quillworts, this species is found throughout the entire eastern United States. In Georgia, it has been found in 12 counties, ranging from Tift and Worth to Rabun and Catoosa.

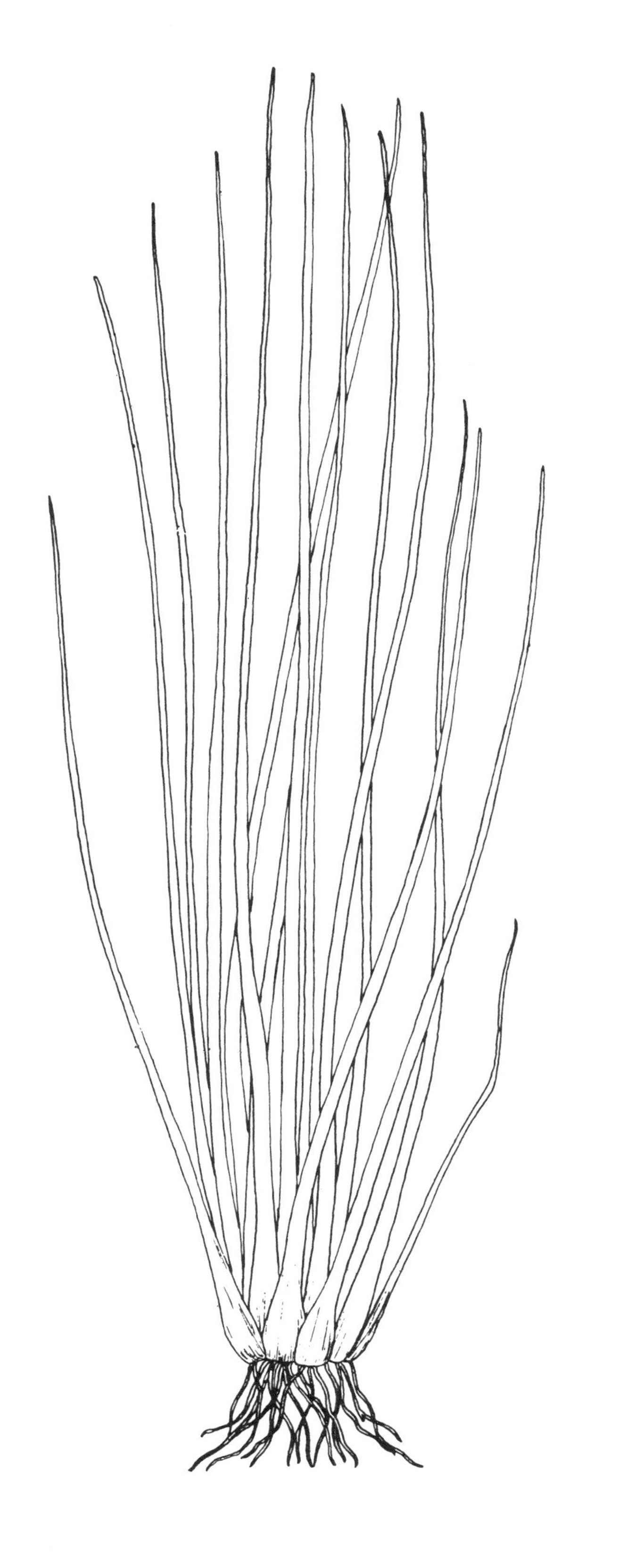

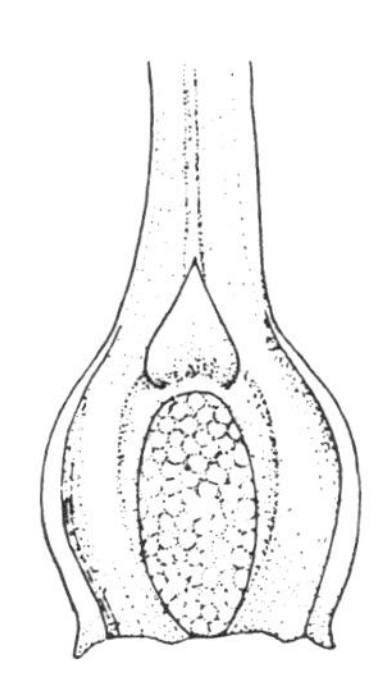

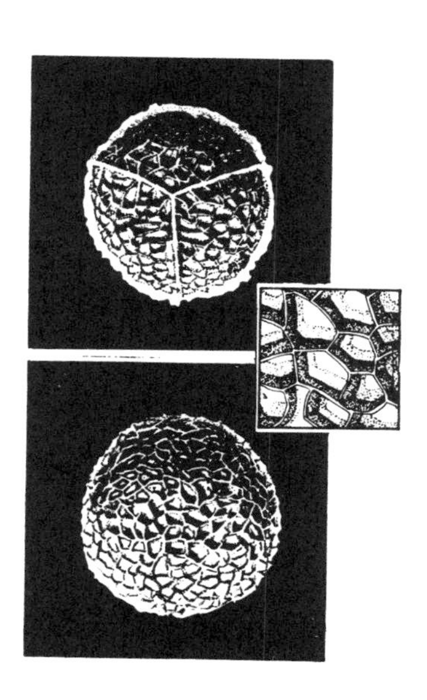

Hybrid Quillworts

Isoetes engelmannii × *piedmontana* and
Isoetes flaccida × *piedmontana*

In 1982 Brian M. Boom determined by laboratory studies that a few of the Georgia specimens of *Isoetes*, previously thought to be species, were actually hybrids.

Isoetes engelmannii × *piedmontana* is characterized by reticulo-cristate megaspores, brown-spotted sporangial walls, and a diminutive sporophyll habit. It has been found in three widely separated Georgia counties.

Isoetes flaccida × *piedmontana* is distinct in having 70 to 90 percent velum coverage and pigmented sporangial walls. It, too, has been found in five widely separated counties.

Isoetes engelmannii × piedmontana

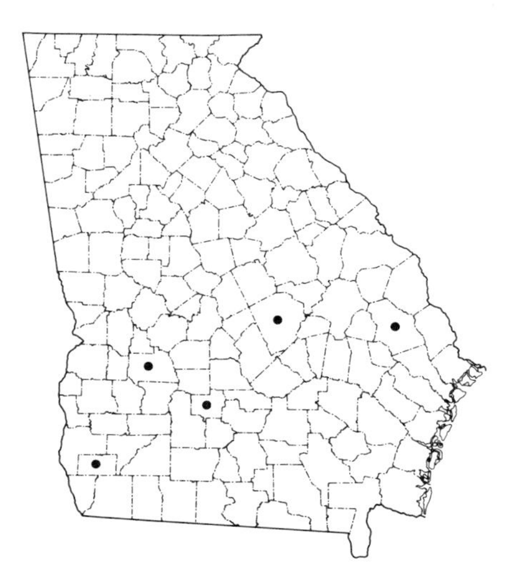

Isoetes flaccida × piedmontana

Black-Spored Quillwort

Isoetes melanospora Engelm.

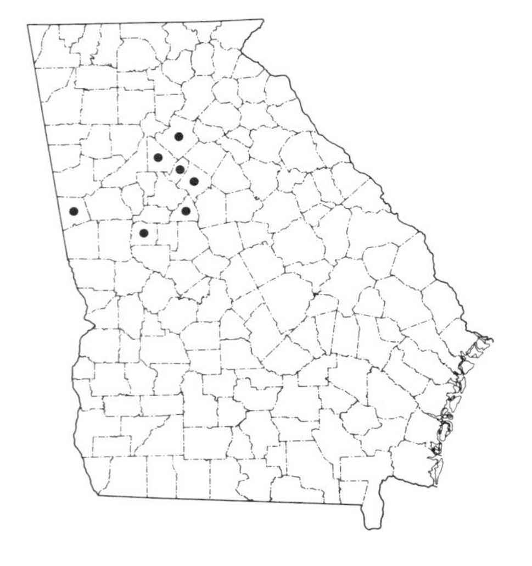

Name: William M. Canby, an amateur botanist, first found this quillwort on Stone Mountain in 1869, and George Engelmann of St. Louis named it in 1877. The common name comes from the Latin epithet.

Roots: Monomorphic, forking. Corm 2-lobed.

Leaves: The smallest of the quillworts, the leaves are slender-tipped, whorled, and only 3 to 8 cm long. 5 to 10 per plant. Stomata present, peripheral strands absent.

Sporangia: Circular. 1 to 2 mm in diameter. Walls without pigmented cells. Completely covered by velum.

Spores: Mature early spring. Megaspores 280 to 400 microns in diameter, black, tuberculate. Microspores 25 to 30 microns long, brown, smoothish.

Habitat: Amphibious in very thin, acidic soil of shallow pools on granite domes and flatrocks.

Range: Granite domes and outcrops of Georgia's central Piedmont and one location in north-central South Carolina.

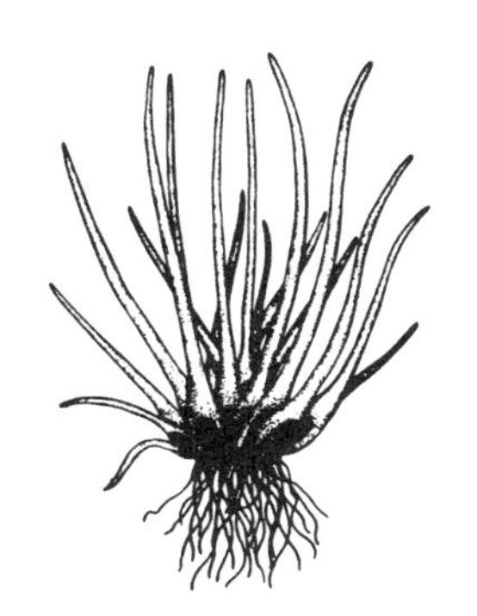

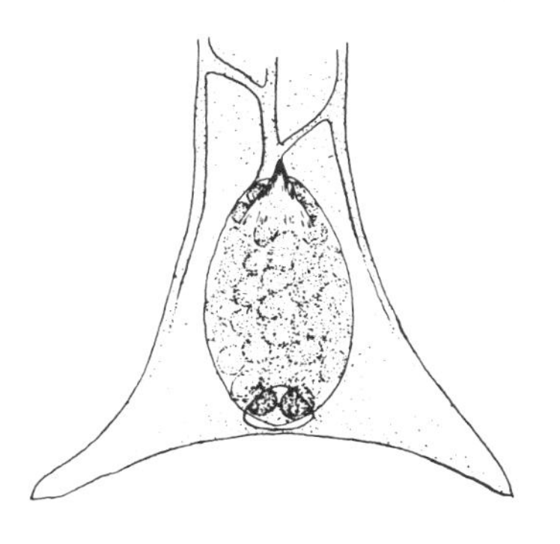

Mat-Forming Quillwort

Isoetes tegetiformans Rury

Name: This new quillwort was discovered by Phillip M. Rury on Heggie's Rock in Columbia County, Georgia, in 1978. Rury named it *tegetiformans* because of its unique mat-forming characteristic that contrasts with the solitary plant forms of the other species of *Isoetes*.

Roots: Frequently dimorphic, with two rows of slender roots and one row of stout, often curled roots. Unforked.

Stems: 3 to 35 mm long, prostrate, often curved and grooved in center. Frequently bear distal, adventitious plantlets, so that plants are dense, mat-forming.

Leaves: 2 to 4 cm long, in groups of 4 to 8. Stomata present, peripheral strands absent.

Sporangia: Elliptical, kidney-shaped, 1 mm long, walls without pigmented cells. Velum completely covers sporangia.

Spores: Megaspores 275 to 370 microns in diameter, brown, low tuberculate. Microspores 26 to 31 microns long, brown, echinate.

Habitat: Amphibious or submerged. Shallow pools on granite outcrops.

Range: Unknown except for four Georgia locations. It is on Georgia's list of rare plants.

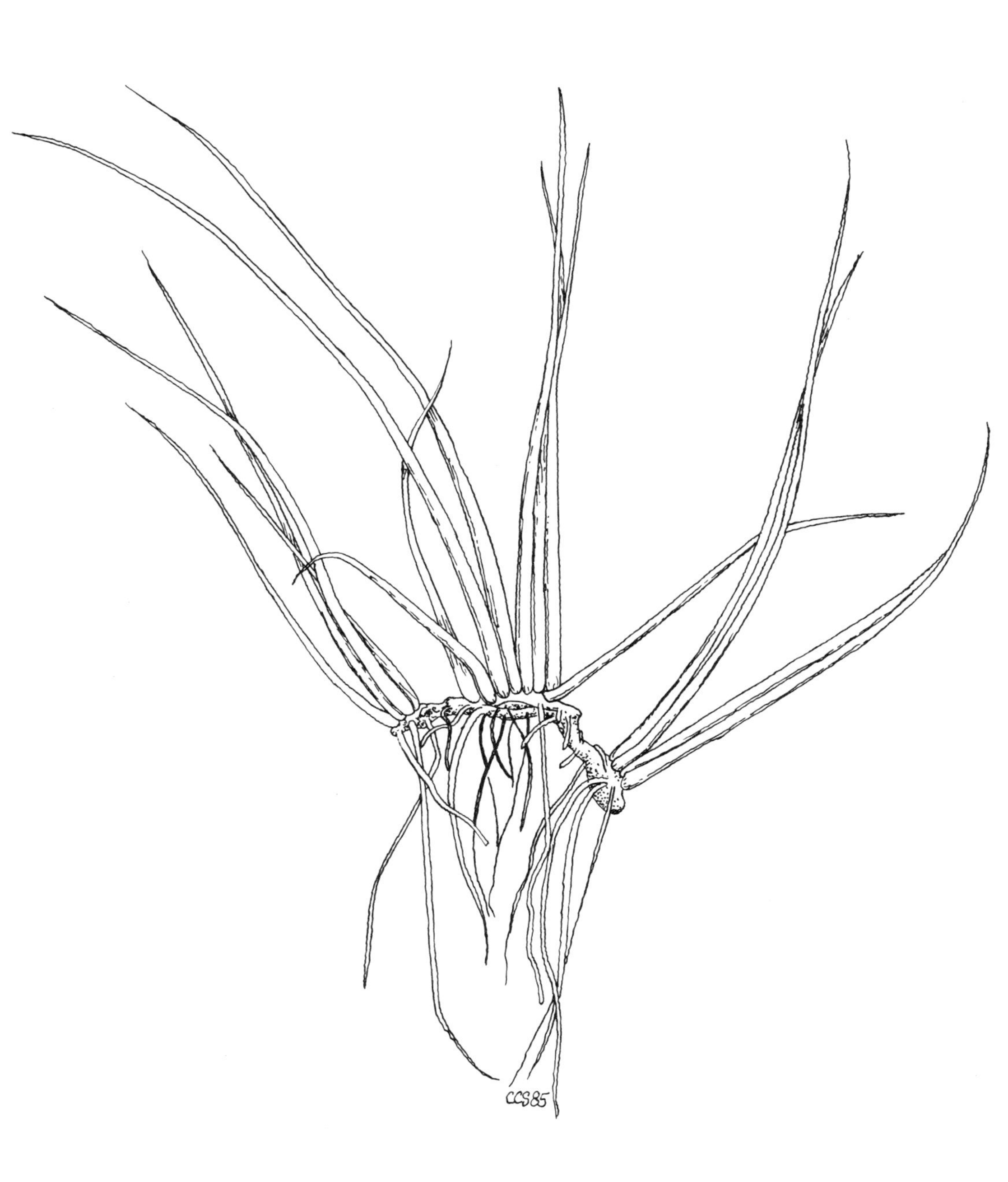
CCS85

Black-Footed Quillwort

Isoetes melanopoda Gay and Dur.

Name: First discovered in the west-central United States, this taxon was named by Gay and Durieu in 1864. The epithet means "black-footed" and refers to the dark-colored bases of the leaves.

Roots: Monomorphic, forking. Corm 2-lobed.

Leaves: 7 to 45 cm long, with shiny black tissue at base. 15 to 60 to each plant. Stomata and peripheral strands present.

Sporangia: Elliptical, 5 to 30 mm long with brown pigmented cells. Velum covers less than one-third.

Spores: Mature in late spring. Megaspores 280 to 440 microns in diameter, light cream–colored or white, tuberculate. Microspores 20 to 30 microns long, brown, echinate.

Habitat: Various. Depressions and ponds, deep soil of eastern prairies or thin soil pockets of sandstone outcrops.

Range: Frequent and widespread in the midwestern United States. Also found in the central and southeastern states. Only location in Georgia is in Floyd County.

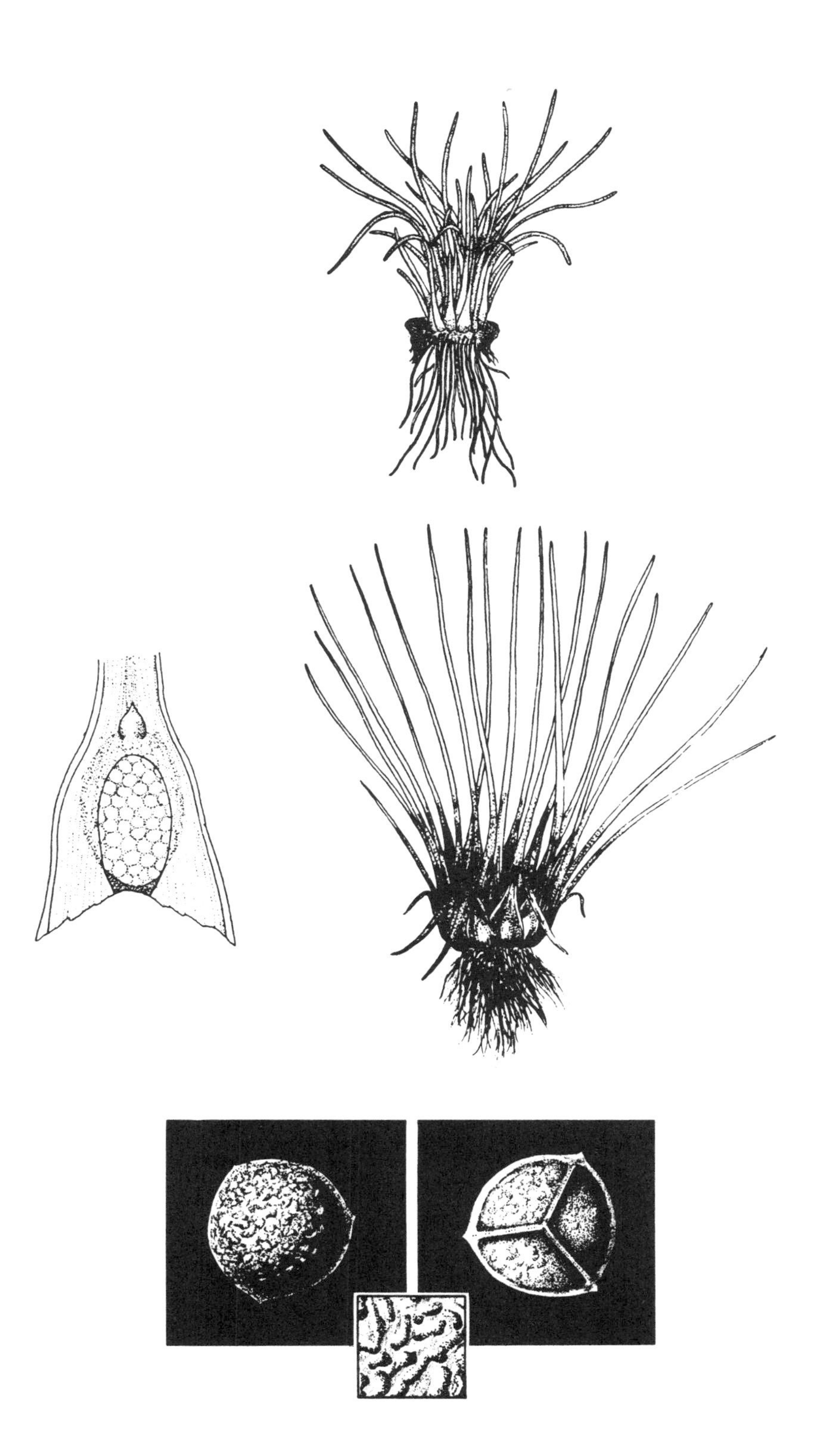

Quillwort

Isoetes butleri Engelm.

Name: In 1878 George Engelmann named this taxon in honor of G. D. Butler.

Roots: Monomorphic, forked. Corm 2-lobed.

Leaves: 8 to 20 cm long. Lax. 8 to 20 in each plant. Stomata numerous and peripheral strands present.

Sporangia: 6 to 14 mm long. Elliptical. Marked with brown lines. Velum covers less than one-third. Walls have pigmented cells.

Spores: Mature in late spring. Megaspores, 480 to 650 microns in diameter, light cream–colored or white, tuberculate. Microspores, 27 to 37 microns long, brown with wartlike surfaces.

Habitat: Moist to wet calcareous soil of cedar glades or barrens.

Range: The southeastern United States. About 8 sites in Tennessee and Arkansas and a few locations in Missouri, Kentucky, and Alabama. In Georgia, only in the Chickamauga National Military Park in Catoosa County.

Quillwort

Isoetes flaccida Shuttlew. ex A. Br.

Name: This species was discovered in northern Florida in 1842. Shuttleworth's name, published by Alexander Braun in 1846, refers to the limpness, or flaccidity, of the leaves.

Roots: Monomorphic, forked. Corm 2-lobed.

Leaves: 10 to 60 cm long. 10 to 35 to a plant. Stomata numerous, peripheral strands occasional.

Sporangia: Elliptical. 4 to 6 mm long. No pigmented cells and completely covered by velum.

Spores: Mature in summer and early autumn. Megaspores, 300 to 540 microns in diameter, cream-colored or white, tuberculate. Microspores, 20 to 33 microns in diameter, light brown, surface papillate (bearing minute, nipplelike projections).

Habitat: Amphibious. Running or standing water in wet, sandy depressions, cypress swamps, and margins of streams and lakes.

Range: Rare. Limited to Florida (mostly western sections) and a few sites in southern Georgia.

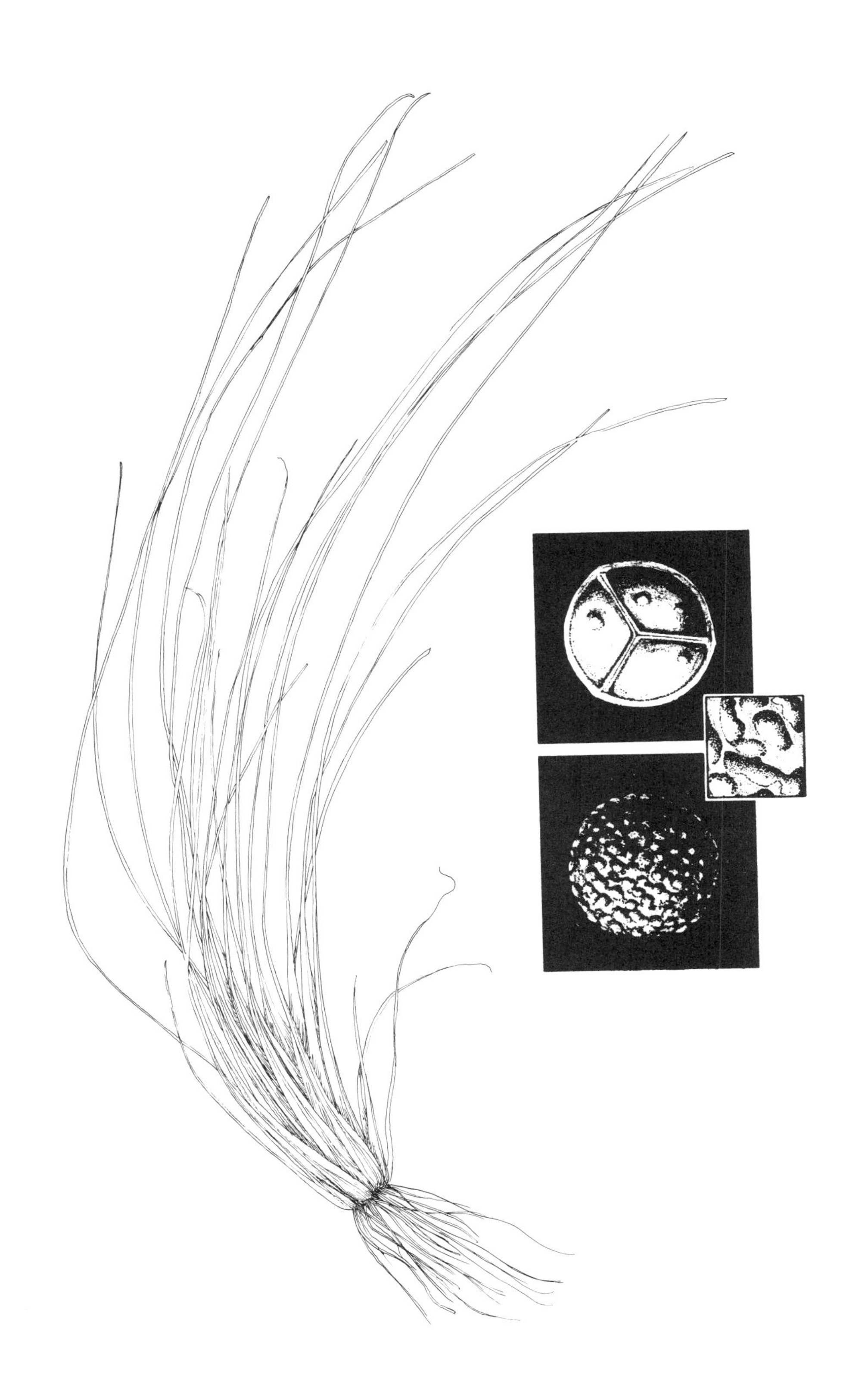

Black-Based Quillwort

Isoetes piedmontana (Pfeiffer) Reed

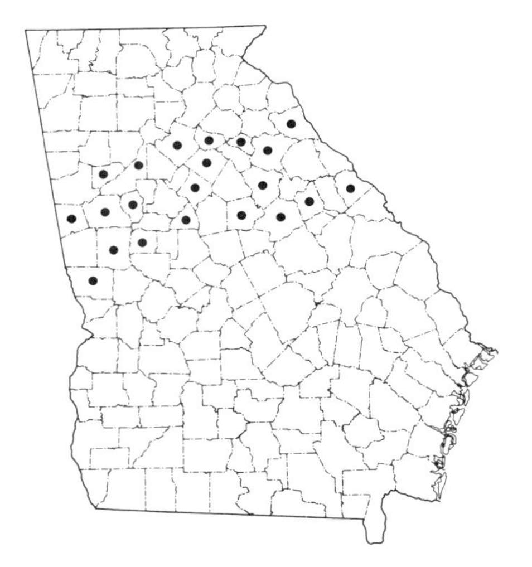

Name: Similar to the rare Virginia quillwort *I. virginica*, Norma E. Pfeiffer named this *I. virginica* var. *piedmontana* in 1939. C. F. Reed elevated it to species rank in 1965. The Latin name comes from the southeastern Piedmont area where it is found, and the common name from the leaf coloring.

Roots: Monomorphic, forked. Corm 2-lobed.

Leaves: 7 to 15 cm long, often brownish at base. 15 to 60 to a plant. Both stomata and peripheral strands present.

Sporangia: Elliptical, 3 to 5 mm long. Walls have pigmented cells, with velum covering less than one-third.

Spores: Mature in early spring. Megaspores 400 to 800 microns in diameter, cream-colored or white, tuberculate. Microspores 27 to 37 microns long, brownish, echinate.

Habitat: Amphibious. Acidic soil in shallow pools on granite domes and flat rocks.

Range: Most commonly found *Isoetes* in Georgia. In the central and southern Piedmont from the east to the west borders of the state. Outside Georgia, in one site each in Alabama and South Carolina.

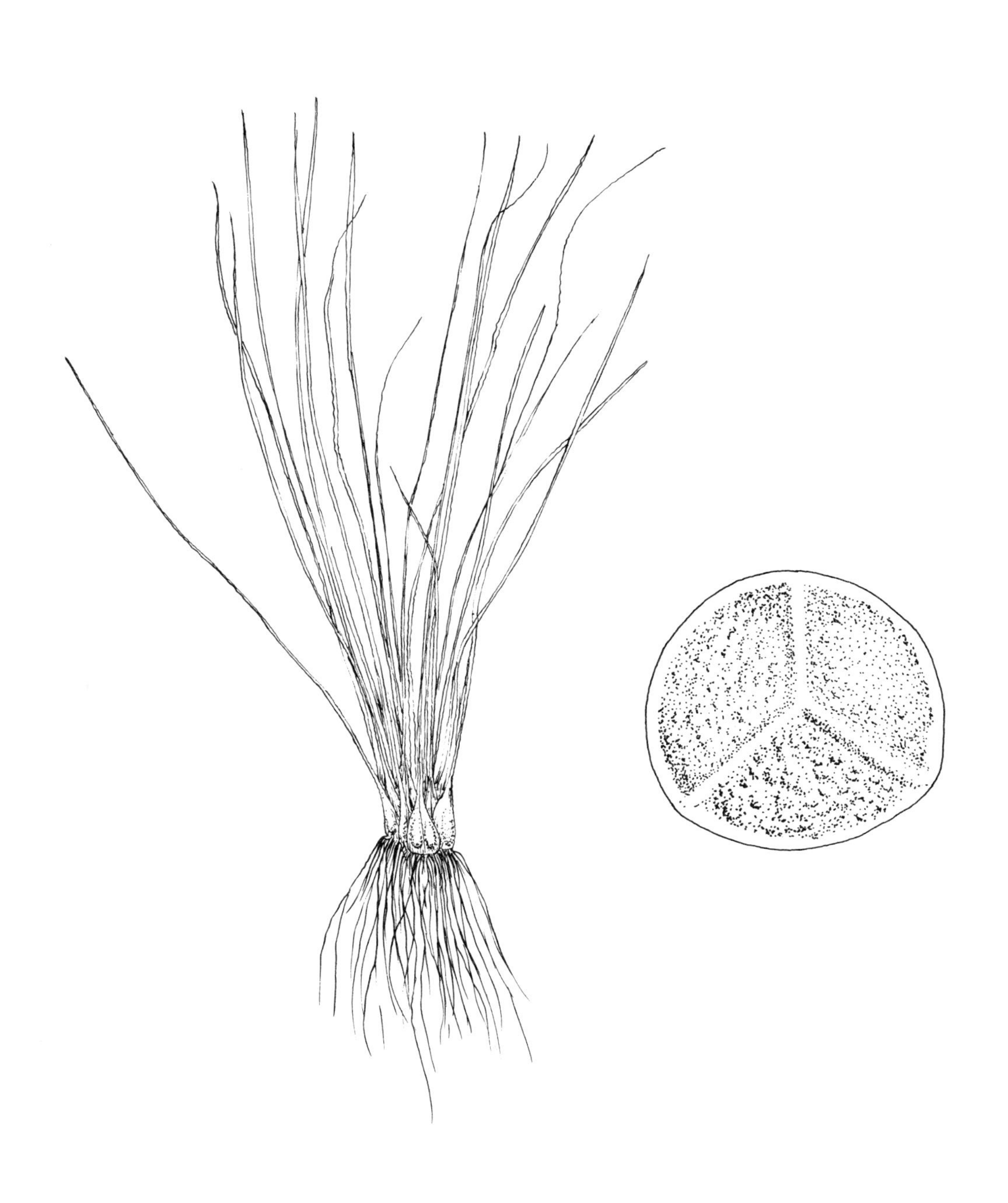

Field Horsetail

Equisetum arvense L.

Name: Linnaeus described this species in 1753, the scientific name meaning almost literally "field horsetail."

Rootstock: Slender, much-branched, creeping. Roots and occasionally tubers at nodes.

Fertile Stems: Annual, unbranched. 15 to 30 cm high. Appear in early spring, shed spores, and soon wither. Pink or tan, succulent. Sheaths with large, dark, lanceolate teeth.

Sterile Stems: Annual. Branched but not rebranched. 25 to 50 cm high. Appear as fertile die. Green. Slender branches in whorls at nodes. Node sheaths slightly flaring, teeth 6 to 12, brown-tipped. Internodes 1 to 5 cm, lowest usually longest.

Canals: Central canal one-half diameter of stem. Vallecular canals about one-third size of central canal.

Strobili: 1 to 3 cm long on short stalk at terminal end of fertile stem.

Habitat: Damp woods, swamps, stream banks, meadows, roadsides, and railroad embankments.

Range: Abundant in North America, Europe, and Asia. Most common North American horsetail, but rare in the southern states. The few sites in northern Georgia are along its southernmost limits.

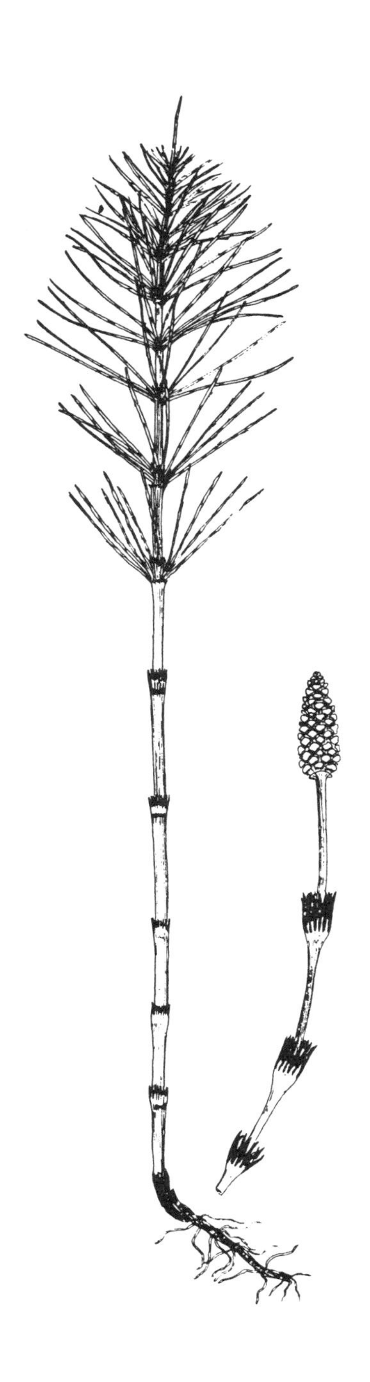

Scouring Rush

Equisetum hyemale L.

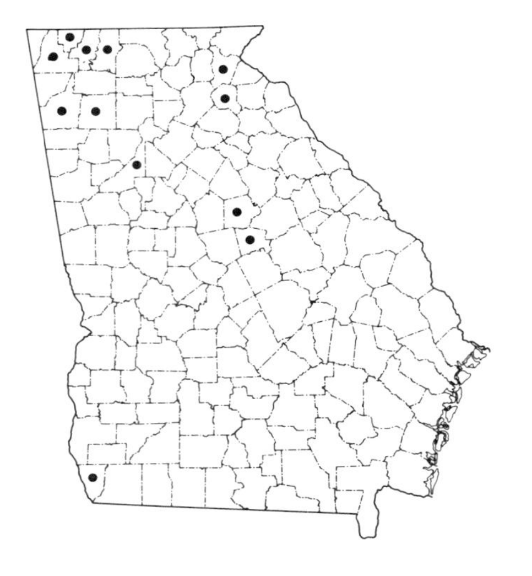

Name: Linnaeus named this taxon in 1753. The epithet means "of winter," referring to its evergreen nature.

Rootstock: Wide-creeping and -branching.

Stems: Fertile and sterile alike. Evergreen, unbranched, bamboolike. 50 to 100 cm tall. Internodes 1 to 8 cm. Ridges 18 to 40 in number. Sheaths cylindrical, lying close against stem. Black band above and below sheath. Teeth narrow, sharp-pointed, dark brown with white margin, persistent or deciduous.

Canals: Central four-fifths diameter of stem. Vallecular canals about one-sixth size of central.

Strobili: Short-stemmed, with sharp, pointed tip.

Habitat: Damp locations including wet places in woods, along watercourses, sandy shores, and wet, shaded slopes. Invades disturbed soil.

Range: Common in Europe, Asia, and almost all of North America. Found in scattered locations in Georgia.

Whisk Fern

Psilotum nudum (L.) Beauv.

Name: Linnaeus first described this pteridophyte in 1753 as *Lycopodium nudum,* the specific name meaning "bare," referring to the green, leafless or naked branches. Palisot de Beauvois transferred it in 1805 to *Psilotum,* a genus established by Swartz in 1801. This name is Greek for "naked" or "smooth," also referring to the leafless branches.

Rhizome: Slender, creeping underground, fleshy, branched, black, hairy, without roots.

Erect Stems: 15 to 50 cm long. Green. Branching several times in nearly equal halves. Tiny, scalelike bracts, few in number and irregularly spaced along stems. Plant appears somewhat like a leafless shrub with green twigs.

Sporangia: Solitary, 3-lobed. Borne on upper parts of stem. Often thicker than branch on which they are borne.

Habitat: Damp woods and swamps, tree bases, logs, hammocks. Also an epiphyte in tree crotches and bark crevices.

Range: Throughout the tropics. Rare in the southwestern and southeastern United States, except sometimes common locally in Florida. In Georgia, has been found in a few southeastern counties of the Lower Coastal Plain.

Remarks: The absence of roots and leaves indicates that *Psilotum* is one of the most unusual and simple of vascular plants.

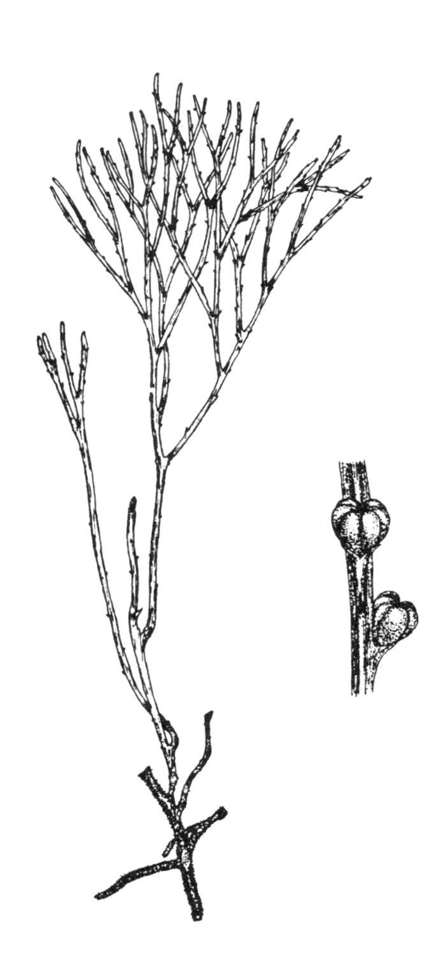

Checklist

FERNS

__ *Adiantum capillus-veneris*
__ *Adiantum pedatum*
__ *Asplenium bradleyi*
__ *Asplenium × ebenoides*
__ *Asplenium × gravesii*
__ *Asplenium heterochroum*
__ *Asplenium heteroresiliens*
__ *Asplenium × kentuckiense*
__ *Asplenium montanum*
__ *Asplenium pinnatifidum*
__ *Asplenium platyneuron*
__ *Asplenium resiliens*
__ *Asplenium ruta-muraria*
__ *Asplenium trichomanes*
__ *Asplenium × trudellii*
__ *Asplenium × wherryi*
__ *Athyrium filix-femina*
__ *Athyrium pycnocarpon*
__ *Athyrium thelypterioides*
__ *Azolla caroliniana*
__ *Azolla filiculoides*
__ *Blechnum occidentale*
__ *Botrychium biternatum*
__ *Botrychium dissectum*
__ *Botrychium jenmanii*
__ *Botrychium lunarioides*
__ *Botrychium virginianum*
__ *Camptosorus rhizophyllus*
__ *Cheilanthes alabamensis*
__ *Cheilanthes lanosa*
__ *Cheilanthes sinuata*
__ *Cheilanthes tomentosa*
__ *Cyrtomium falcatum*
__ *Cyrtomium fortunei*

___ *Cystopteris bulbifera*
___ *Cystopteris protrusa*
___ *Cystopteris tennesseensis*
___ *Dennstaedtia punctilobula*
___ *Diplazium japonicum*
___ *Dryopteris × australis*
___ *Dryopteris celsa*
___ *Dryopteris cristata*
___ *Dryopteris goldiana*
___ *Dryopteris intermedia*
___ *Dryopteris ludoviciana*
___ *Dryopteris marginalis*
___ *Dryopteris × neo-wherryi*
___ *Lorinseria areolata*
___ *Lygodium japonicum*
___ *Lygodium palmatum*
___ *Onoclea sensibilis*
___ *Ophioglossum crotalophoroides*
___ *Ophioglossum engelmannii*
___ *Ophioglossum nudicaule*
___ *Ophioglossum petiolatum*
___ *Ophioglossum pycnostichum*
___ *Osmunda cinnamomea*
___ *Osmunda claytoniana*
___ *Osmunda regalis*
___ *Pellaea atropurpurea*
___ *Pilularia americana*
___ *Polypodium aureum*
___ *Polypodium polypodioides*
___ *Polypodium virginianum*
___ *Polystichum acrostichoides*
___ *Pteridium aquilinum*
___ *Pteris multifida*
___ *Pteris vittata*
___ *Salvinia minima*
___ *Thelypteris dentata*
___ *Thelypteris hexagonoptera*
___ *Thelypteris hispidula*
___ *Thelypteris kunthii*
___ *Thelypteris noveboracensis*

___ *Thelypteris ovata*
___ *Thelypteris palustris*
___ *Thelypteris torresiana*
___ *Trichomanes boschianum*
___ *Trichomanes petersii*
___ *Vittaria lineata*
___ *Woodsia obtusa*
___ *Woodsia scopulina*
___ *Woodwardia virginica*

OTHER PTERIDOPHYTES

___ *Equisetum arvense*
___ *Equisetum hyemale*
___ *Isoetes butleri*
___ *Isoetes engelmannii*
___ *Isoetes engelmannii × piedmontana*
___ *Isoetes flaccida*
___ *Isoetes flaccida × piedmontana*
___ *Isoetes melanopoda*
___ *Isoetes melanospora*
___ *Isoetes piedmontana*
___ *Isoetes tegetiformans*
___ *Lycopodium alopecuroides*
___ *Lycopodium alopecuroides × appressum*
___ *Lycopodium alopecuroides × prostratum*
___ *Lycopodium appressum*
___ *Lycopodium appressum × prostratum*
___ *Lycopodium carolinianum*
___ *Lycopodium cernuum*
___ *Lycopodium clavatum*
___ *Lycopodium digitatum*
___ *Lycopodium lucidulum*
___ *Lycopodium obscurum*
___ *Lycopodium prostratum*
___ *Lycopodium selago*
___ *Lycopodium tristachyum*
___ *Psilotum nudum*
___ *Selaginella acanthonota*
___ *Selaginella apoda*

___ *Selaginella arenicola*
___ *Selaginella braunii*
___ *Selaginella kraussiana*
___ *Selaginella ludoviciana*
___ *Selaginella riddellii*
___ *Selaginella rupestris*
___ *Selaginella tortipila*
___ *Selaginella uncinata*

Glossary

Acuminate. Tapering to a point.
Alternate. Arranged at different heights; not opposite.
Anterior. On the front side.
Apex. The tip or end.
Appressed. Lying flat, or pressed against the surface.
Areole. The space surrounded by leaf veins.
Auricle. An earlike lobe.
Axil. The angle formed by the axis and a leaf.
Axis. A main stem or stalk.

Blade. The broad, flat portion of a frond or leaf.
Bract. A small or modified leaf.
Bulbous. Bulblike.

Calcareous. Containing considerable lime; alkaline.
Caudate. Bearing a tail-like appendage.
Circumneutral. Only very slightly acid or alkaline.
Cone. A cylindrical group of sporophylls.
Connivent. Coming together or converging but not fused.
Cordate. Heart-shaped.
Corm. Fleshy, bulblike base of a stem.

Deciduous. Not evergreen.
Decurrent. Extending downward.
Dimorphic. Of two kinds.
Dissected. Cut or divided into numerous segments.
Dorsal. The back or lower surface or side.

Elliptic. Rounded in outline, but broadest in middle.
Entire. Margin continuous, not cut or indented.
Epiphyte. A plant growing on another for support.
Epithet. Part of a scientific name following the genus and together with the genus designating a species.

Fertile. Producing spores.
Frond. A fern leaf including blade and stipe.

Gametophyte. The sexual generation of a fern; the prothallus.
Gemma. A vegetative reproductive bud.
Genus. A natural group of closely related species.
Glabrous. Smooth; without hairs or scales.
Globose. Spherical or nearly so.
Glochidia. Anchor-shaped hairs among microspores of *Azolla*.

Heterosporous. Having two kinds of spores.
Hirsute. Hairy.
Homosporous. Having only one kind of spores.
Hybrid. Plant resulting from the crossing of two species.

Incised. Cut irregularly, more or less deeply.
Indusium. A membrane covering a fern sorus.
Inferior. Lying beneath or on the lower side.
Internodes. The part of a stem between two nodes.

Lanceolate. Lance-shaped. Broadest between base and middle and tapering toward apex.

Lateral. Lying at the side.
Linear. Long, narrow, with sides nearly parallel.
Lobe. A short, rounded division of a blade.

Marginal. At the margin or edge of a blade or blade division.
Medial. In the middle portion.
Megaspore. The large, female spore of *Azolla, Pilularia, Isoetes, Selaginella,* and *Salvinia.*
Microspore. The small, male spore of *Azolla, Pilularia, Isoetes, Selaginella,* and *Salvinia.*
Midrib. The central vein of a leaf or leaf division.
Monomorphic. Having only one form.

Node. Point on a stem where one or more leaves are attached.

Oblanceolate. Lanceolate with broadest part toward tip instead of base.
Opposite. Said of leaves attached to opposite sides of stem at same node.
Orbicular. Circular or nearly so.
Ovate. Egg-shaped, broadest below middle.

Palmate. Divided radiately like the fingers of a hand.
Peduncle. The stalk of a cone of a *Lycopodium.*
Peltate. Supported on a central stalk, like an umbrella.
Persistent. Living a long time; surviving winters.
Petiole. The stalk of a leaf.
Pinna. A primary division of a fern frond that is narrowed at the base.
Pinnate. Divided with divisions narrowly attached to the rachis.
Pinnatifid. Divided with cuts not going all the way to the midrib.
Pinnule. A secondary division of a fern frond; a division of a pinna that is narrowed at the base.
Prostrate. Lying flat.
Prothallus. The small plant constituting the fern gametophyte.
Pteridophytes. Vascular plants reproduced by spores.
Pubescent. Hairy.

Rachis. The axis or midrib of a compound fern blade.
Reflexed. Abruptly bent downward or backward.
Reniform. Kidney-shaped.
Rhizome. A horizontal creeping stem.
Rootstock. Prostrate or underground stem.

Scale. Thin chaff on fern parts.
Segment. Division of a frond blade that is deeply lobed but not cut all the way to the axis or midrib.
Serrate. With edges cut into sawlike teeth.
Sessile. Without a stalk.
Seta. A hairlike extension of a leaf margin.
Sinus. The notch or gap between lobes or segments.
Sorus. A cluster of sporangia.
Species. Basic unit of plant classification.
Sporangium. A spore case.
Spore. A single-cell reproductive body.
Sporocarp. A podlike body containing spores of heterosporous pteridophytes.
Sporophyll. A fertile leaf bearing sporangia and spores.
Sporophyte. A plant producing spores.
Stipe. The stalk supporting the blade of a fern frond.
Stomata. Minute breathing pores on a leaf surface.
Strobilus. The cone or cluster of fertile leaves on certain pteridophytes.

Subspecies. A major subdivision of a species.
Sulcate. Grooved or furrowed longitudinally.
Superior. On the upper side.

Ternate. In threes.

Variety. A lesser subdivision of a species.
Vascular. Having conducting vessels or tissues running through it.
Velum. A tissue covering, in whole or part, the sporangium in *Isoetes.*

References

BOOKS

Brown, C. A., and D. S. Correll. Ferns and Fern Allies of Louisiana. Baton Rouge: Louisiana State University Press, 1942.

Bruce, J. G. Systematics and Morphology of Subgenus *Lepidotus* of the Genus *Lycopodium* (Lycopodiaceae). Ph.D. dissertation, University of Michigan, 1975.

Cobb, Boughton. A Field Guide to the Ferns. Boston: Houghton Mifflin Company, 1956.

Cranfill, Ray. Ferns and Fern Allies of Kentucky. Frankfort: Kentucky Nature Preserves Commission, 1980.

Key, James S. Field Guide to Missouri Ferns. Jefferson City: Missouri Department of Conservation, 1982.

Lakela, O., and R. W. Long. Ferns of Florida. Miami: Banyan Books, 1976.

McVaugh, Rogers, and Joseph H. Pyron. Ferns of Georgia. Athens: University of Georgia Press, 1951.

Mickel, John T. How to Know the Ferns and Fern Allies. Dubuque, Iowa: Wm. C. Brown Company, 1979.

Shaver, Jesse M. Ferns of Tennessee. Nashville: George Peabody College for Teachers, 1954.

Small, J. K. Ferns of the Southeastern States. Lancaster, Pa.: Science Press, 1938.

Smith, A. R. Systematics of Neotropical Species of *Thelypteris* (section *Cyclosorus*). Vol. 59. Berkeley: University of California Publications in Botany, 1971.

Wherry, Edgar T. The Fern Guide. Garden City: Doubleday and Company. 1961. Reprint. Philadelphia: Morris Arboretum, 1975.

———. The Southern Fern Guide. Garden City: Doubleday and Company, 1964. Reprint. New York: New York Chapter of the American Fern Society, 1977.

ARTICLES

Beitel, Joseph. Clubmosses (*Lycopodium*) in North America. *Fiddlehead Forum* 6(5):1–8. 1979.

Boom, Brian M. Synopsis of *Isoetes* in the southeastern United States. *Castanea* 47:38–59. 1982.

Bruce, James G., Samuel B. Jones, and Nancy C. Coile. The pteridophytes of Georgia. *Castanea* 45:185–93. 1980.

Clausen, Robert T. *Selaginella,* subgenus *Euselaginalla,* in the southeastern United States. *American Fern Journal* 36:65–82. 1946.

Coile, Nancy C. Flora of Elbert County, Georgia. *Castanea* 46:173–94. 1981.

Duncan, Wilbur H. New records for Georgia ferns. *American Fern Journal* 45:1–10. 1955.

Faircloth, Wayne R. Ferns and other primitive vascular plants of central south Georgia. *Castanea* 40:217–28. 1975.

Faircloth, Wayne R., and Mary Norsworthy. Notes on some *Ophioglossaceae* in the Georgia

coastal plain. *American Fern Journal* 65:28. 1975.
———. *Diplazium japonicum* and *Selaginella uncinata* newly discovered in Georgia. *American Fern Journal* 71:48–50. 1981.
Jones, Samuel B. Flora and phytogeography of the Pine Mountain region of Georgia. *Castanea* 39:113–49. 1974.
Montgomery, James D. *Dryopteris* in North America Part II: The hybrids. *Fiddlehead Forum* 9(4):23–30. 1982.
Montgomery, James D., and E. M. Paulton. *Dryopteris* in North America. *Fiddlehead Forum* 8(4):25–31. 1981.
Norsworthy, Juanita. Fern records for Echols County and the state of Georgia. *American Fern Journal* 56:55–57. 1966.
Rury, Phillip M. A new and unique mat-forming Merlin's grass (*Isoetes*) from Georgia. *American Fern Journal* 66:99–107. 1978.
Smith, Dale M., Truman R. Bryant, and Donald E. Tate. Another *Asplenium* hybrid from Kentucky. *American Fern Journal* 51:70–73. 1961.
Van Eseltine, G. P. The allies of *Selaginella rupestris* in the southeastern United States. *Contributions from the National Herbarium.* 20:159–72. 1918.
Wagner, W. H., Jr. Two new species of ferns from the United States. *American Fern Journal* 56:12–16. 1966.
———. Reticulate evolution in the Appalachian *Aspleniums. Evolution* 8:103–18. 1954.
Wagner, W. H., Jr., Donald R. Farrar, and Bruce W. McAlpin. Pteridology of the Highlands Biological Station area, Southern Appalachians. *Journal of the Elisha Mitchell Scientific Society* 86:1–17. 1970.

Index

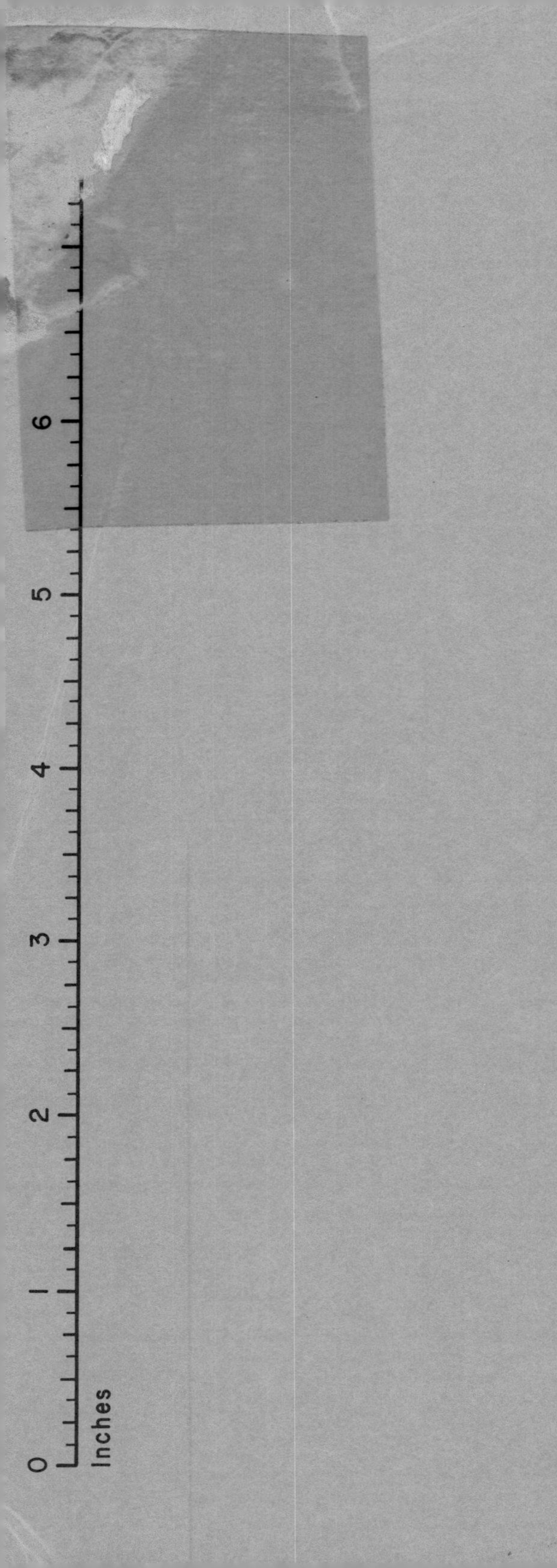